Urban Rivers Our Inheritance and Future

Edited by **Geoff Petts** University of Birmingham **John Heathcote** Entec UK Ltd **Dave Martin** Environment Agency

Published by IWA Publishing, Alliance House, 12 Caxton Street, London SW1H 0QS, UK
Telephone: +44 (0) 20 7654 5500; Fax: +44 (0) 20 7654 5555; Email: publications@iwap.co.uk
Web: www.iwapublishing.com

First published 2002

Printed by Henry Ling Ltd, Dorchester, UK
Designed by Smith & Gilmour

British Library Cataloguing in Publication Data
A CIP catalogue record for this book is available from the British Library

Library of Congress Cataloging- in-Publication Data
A catalog record for this book is available from the Library of Congress

ISBN: 1 900222 22 1
Environment Agency R&D p2–156.

Acknowledgements
This book has been a team effort. We are delighted to acknowledge the support of Lynne Pearce (University of Birmingham) in coordinating editorial activities and in undertaking background research across a range of areas, and our graphic artist Neil Patton (Entec) for all of the drawings. Valuable technical contributions have been provided by:

Nigel Brown (University of Birmingham)
Martin Carville (Entec)
Kevin Chipman (Universtiy of Birmingham)
Angela Davenport (University of Birmingham)
George Dobson (Entec)
Chris Doyle (University of Birmingham)
Garry Edwards (Entec)
John Hall (Entec)
Nick Morton (University of Birmingham)
Judith Petts (University of Birmingham)
John Pomfret (Entec).

We are also grateful for the input received from over 30 of the Environment Agency's and SEPA's technical experts and the design team at IWA Publishing. Finally, we would like to acknowledge the work of John Tyson who helped initiate the project that produced this book.

CONTENTS

The Environment Agency

The Environment Agency was established by the 1995 Environment Act and is one of the largest and most powerful environmental protection agencies in Europe. The Agency has wide-ranging responsibilities and strong powers to protect and, where necessary, improve the environment in England and Wales.

In carrying out its work, the Agency is guided by its duty to protect the environment in a way that works towards achieving sustainable development. This involves meeting the needs of present generations without compromising the ability of future generations to meet their own needs.

Scottish Environment Protection Agency (SEPA)

The Scottish Environment Protection Agency (SEPA) is the public body responsible for environmental protection in Scotland. Its main aim is "to provide an efficient and integrated environmental protection system for Scotland which will both improve the environmental and contribute to the Government's goal of sustainable development."

SEPA regulates discharges to waters and the air, the storage, transport and disposal of controlled waste and the keeping and disposal of radioactive materials. SEPA provides extensive guidance and advice to regulated organisations and works in partnership with others to deliver environmental goals through non-statutory means.

SNIFFER

Scotland and Northern Ireland Forum for Environmental Research (SNIFFER) is a company and registered charity that commissions scientific research into water, air, waste and the environment on behalf of its members (Scottish Environment Protection Agency, Environment & Heritage Service Northern Ireland, The Scottish Executive, Scottish Natural Heritage and The Forestry Commission). All research outputs are in the public domain.

SNIFFER's mission is to contribute to the improvement of the environment of Scotland and Northern Ireland, through coordinated dialogue and environmental research, by taking account both of the business needs of its members and the interests of other relevant stakeholders.

IWA

IWA Publishing is the wholly owned subsidiary of the International Water Association (IWA) providing information services on all aspects of water, wastewater and related environmental fields.

IWA Publishing is uniquely placed to respond to the information needs of the international water industry and research community. It offers a high-quality, cost-effective publishing service with an unrivalled ability to promote its publications to the international water community through IWA's worldwide membership.

Entec UK Ltd

Entec UK Ltd (part of the international Suez group) is a professional consultancy firm providing advice and solutions for a wide range of environmental issues. Entec plans new developments such as housing estates and quarries, to minimise their negative effects on the environment so that they contribute to a sustainable future. Entec assists existing businesses in improving their operations so that less waste is produced, environmental impacts are reduced, and the requirements of increasingly tough environmental and safety legislation are met. Lastly, Entec works with major companies and central government to clean up the legacy of contamination resulting from past activities.

University of Birmingham

The University of Birmingham has a national and international reputation for excellence in research and teaching in environmental science, engineering and policy. Around 150 researchers are actively investigating the scientific, technical and socio-economic aspects across a broad range of environmental disciplines, including the management of freshwater resources, environmental restoration, pollution control, environmental risk and human health. The Centre for Environmental Research and Training plays a key role in providing a focus for the University's environmental expertise. It acts as a gateway for external organisations enabling this expertise to be made more widely accessible and provides a mechanism for the promotion and management of interdisciplinary research.

FOREWORD

In his milestone report *Towards an Urban Renaissance*, Lord Rogers set out a vision for urban regeneration, founded on the principles of design excellence, social well-being and environmental responsibility. He made the point that if we fail to promote urban living within the context of a sustainable urban environment, we will increase social deprivation, accelerate the depletion of natural resources and damage biodiversity.

In recent years the water quality of many of our urban streams and rivers has improved dramatically. They can no longer be regarded as little more than open sewers. As a result, urban rivers now represent a significant opportunity for change and can act as a focal point or catalyst for urban regeneration.

Organisations such as the Environment Agency, SEPA and local authorities can make a significant contribution to the management of the urban environment. However, only through partnerships with the private sector and the positive involvement of local communities can we hope to achieve our goal of a renaissance in urban living.

The conflicting pressures on our rivers, particularly those in existing and new urban areas, are considerable. Only by recognising and taking account of these conflicts will we manage our rivers on a sustainable basis. This book aims to identify these pressures and, by drawing on examples of both good and bad practice, seeks to illustrate how urban rivers can be managed to the benefit of the whole of society.

Sir John Harman Chairman
Environment Agency

1 Problems and Opportunities

OUR INHERITANCE

Rivers have always been at the heart of city life; the control of their waters was a key to the building of human societies. The great civilisations that developed in the dry lands around the Mediterranean Sea, and in the Middle East, the Indian subcontinent and China, were founded on flood-based agriculture, and thus depended on rivers.

Across Europe, the history of urban living dates back some 3 000 years, when early towns developed as focal points for trade along inland waterways. Rivers served as an important means of transport at a time when the road network was rudimentary. Towns were established controlling important river crossings, and rivers also provided strategic defences in troubled times.

INDUSTRIALISATION

In the United Kingdom, the stability and prosperity of the Tudor government in the 16th century began an inexorable growth of urban areas. Over the next 300 years, Britain became the workshop of the world, with over half of its population located in towns. The Industrial Revolution saw rivers assuming an ever-widening role, for power, navigation, water supply and waste disposal. With urban and economic growth came wetland drainage for agriculture and urban development, canalisation of natural channels for boat traffic, and embanking and dredging of rivers for flood

Bruges takes its name from the Old Norse "Bryggia" or landing stage and was an early centre for trade. Access to the sea ensured the town's growth and by the Middle Ages it had become the most important commercial centre in north-west Europe. Today the unspoilt historic city attracts millions of visitors each year.

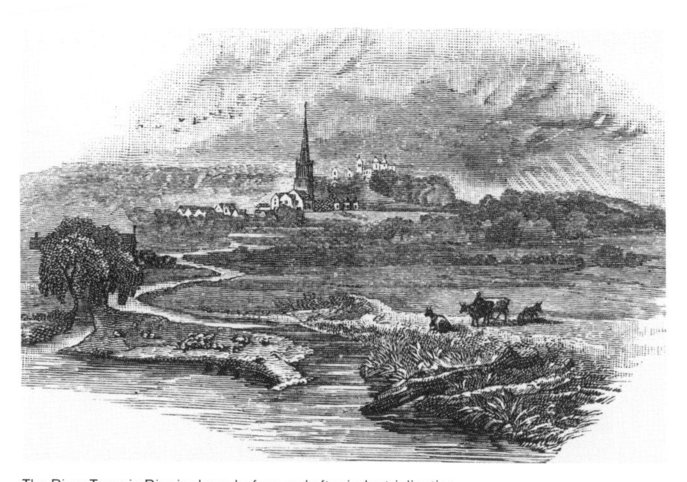

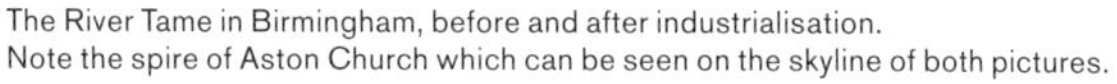

The River Tame in Birmingham, before and after industrialisation.
Note the spire of Aston Church which can be seen on the skyline of both pictures.

control. Freshwater springs dried up as groundwater abstraction increased to supply rapidly growing populations. Overcrowding resulted from the desire to house as many workers as possible as close as possible to the industrial workplace. Ditches and streams became open sewers and breeding grounds for disease. Small streams and brooks were often culverted and became lost, buried beneath expanding urban developments. Large lowland rivers stagnated and important fisheries died. By the 1840s, chronically poor conditions had been reached on many urban streams and rivers.

The environmental legacy of the Industrial Revolution was pollution, slums, and a loss of ownership of urban places and spaces once seen to be at the heart of civilised society. Finally, the change in industrial base during the second half of the 20th century was to lead to the death of the waterfront. Road, rail and air transport replaced river and canal navigation, and a technological revolution caused the virtual disappearance of labour-intensive heavy industry. The demise of the powerful shipbuilding industry of the great deep-water cities of the Thames, Mersey, Tyne and Clyde led to mass unemployment and the decay of riverside urban fabrics.

A CHANGE OF HEART

From a political perspective, one book had a dramatic effect on the public response to the legacy of industrialisation and urbanisation. This was Rachel Carson's *Silent Spring*, published in 1962. Contamination of water and threats to human health and wildlife became major concerns; popular protest movements provided the catalyst for change in policy and practice. Slowly a new vision has evolved, integrating pollution control, the restoration of rivers for wildlife, and improvements for amenity and recreation. Advances in science and technology have now created opportunities for this vision to be realised.

THE FUTURE

During the final years of the last century, the collective endeavours of central Government, local authorities, water companies and developers in the UK have revived the waterfront as a desirable place to live and work. Water frontages provide focal points for new developments and river corridors, even along small streams, can be the arteries for transforming entire urban conurbations.

The challenge for urban development – whether in the UK, elsewhere in the developed world or in developing countries – is to create the quality of life and vitality that makes urban living desirable. Urban revival involves promoting the intimate link between community and nature. Along with

The Greenwich Millennium Village, a showcase for waterfront regeneration. (Courtesy of English Partnerships)

improvements in air quality, building stock and transport networks, green spaces and clean water can be potent symbols of the prosperity of a city.

Using modern technology, rivers can again become the life-blood of cities. They can provide water supplies, hydro-electric power, disposal for treated waste waters, flood channels, and opportunities for navigation, recreation, amenity, fisheries and nature conservation. They can be used to connect different urban areas and to contribute to improving social cohesion, by providing safe green corridors through urban landscapes with visually pleasing clean rivers. Everyone has a part to play in developing and maintaining these local assets.

Safe, clean urban rivers provide numerous opportunities for recreation and enjoyment (River Wye, Bakewell).

HOW TO USE THIS BOOK

This book lays the foundation for everyone to contribute to the rejuvenation of urban streams and rivers. It begins by illustrating the qualities of natural streams and rivers, since we need to understand nature in order to take best advantage of the opportunities provided by urban watercourses. It shows how streams in once rural areas have been transformed by urban development, which has often led to environmental degradation that limits the options available for redevelopment.

The book then focuses on the ways in which modern society uses rivers and the particular problems these uses generate for rivers in today's urban areas. Developments in science and technology are described which provide opportunities to solve these problems.

The book investigates the social, legal and administrative opportunities for finding an integrated approach to reviving urban watercourses. We need to involve everyone in urban renewal: individual members of the public, local communities, planners and developers, water companies, environment agencies and central Government policy makers. Only in this way will we make best use of new opportunities for urban streams and rivers to become the focus for plans to transform urban environments.

Throughout the book, boxed features provide supplementary material, presenting examples and giving more detailed explanations of some concepts.

BOX 1.1 **The global metropolis**

The city is a powerful symbol of social and economic advancement in the history of most nations. Urban development, however, also leads to degradation of basic resources – pollution of rivers, the loss of biological diversity, flooding and flood damage. Development has been founded on a short-term vision driven by a desire to control nature and to exploit natural resources. Many early civilisations collapsed in the face of floods, drought, famine and plague because of their effect on the environment. More recently, the social response to urban living has been one where people have become increasingly alienated from, and indifferent to, the local natural environment. Nature and wild(er)ness have become associated with remote areas, to be visited when desired, for recreation and leisure.

Today urban areas cover 1 per cent of the Earth's surface. The process of urbanisation is accelerating faster than at any other time in history. In 1990 most of the world's population still lived in rural areas, but by 2020 60 per cent – some 4 billion people – are likely to live in cities. Most of this growth is expected to take place in about 30 emerging "megacities", having populations of more than 8 million. (London had a population of around 7 million in 1999.) Most of these are in developing countries where 150 000 people become city dwellers every day. In the UK, some 70 per cent of the population lives in urban areas and nearly 50 per cent lives in cities of over 250 000 people. Urban areas generate 91 per cent of the economic output and provide 89 per cent of the jobs. Urban areas cover over 500 000 hectares (around 2 per cent of the land area) but the road and rail networks extend the urban influence well beyond the limits of cities and conurbations. By 2016 the number of households is likely to have increased by about 4 million. Current Government policy is to locate most of these in large towns and cities.

Street lighting visible from space shows the extent of urban areas across the UK. (NRSC Ltd / Science Photo Library)

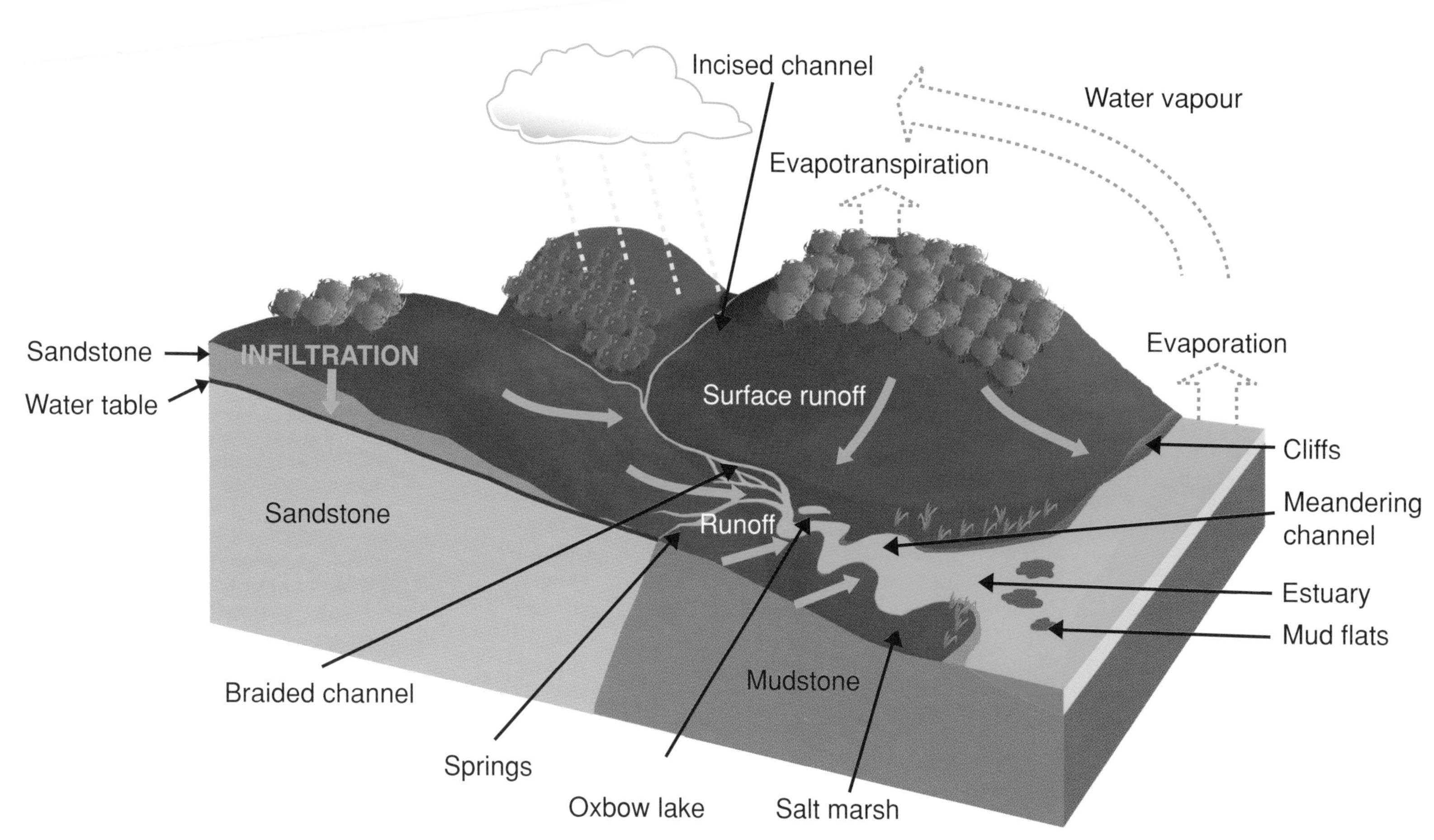

Features and processes of a typical river system, before extensive intervention by man.

2 Setting the Scene

NATURAL RIVERS

Streams and rivers vary in character from place to place: from headwaters to mouth and from region to region across the country. They also change over time, rapidly during floods and gradually over much longer periods. These phenomena are part of urban as much as natural rivers, and must be understood if we are to achieve and sustain urban renewal.

In their natural state, river corridors offer many benefits. These include clean water, productive fisheries, a diverse range of plants and wildlife in and out of the water, navigation routes, flood storage reservoirs (flood plains), and – with great care – a means of waste disposal. But as rivers and their valleys become altered through urbanisation, some of these benefits are lost and others become degraded and more costly to sustain.

The natural water cycle

The water in a river is part of an important natural cycle. This hydrological cycle begins with falling rain and snow that eventually drains into streams. Some water runs off quickly over the surface of the ground, while the remainder soaks in, replenishing groundwater beneath the water table. Groundwater moves slowly, and it may be months or years before it seeps back into rivers or discharges from springs. Streams join to form large rivers, which flow to the seas and oceans from which water evaporates back into the atmosphere. Not all rain drains into rivers; some evaporates from the land surface, lakes and other water bodies, or is used by plants. This process is especially important in summer, when the amount of water lost to the atmosphere from the land exceeds rainfall across most of the UK, leading to drying of the soil and an absence of runoff. During these times, the slow movement of groundwater is essential for maintaining river flows.

Trees play an important role in determining the shape and course of rivers. Woody debris jams provide complex and diverse wildlife habitats.

A catchment is an area of land whose drainage flows into a river system. As rainwater drains from the catchment, it transports sediment, organic matter, nutrients and other dissolved substances into the streams. Human activities within the catchment affect these processes. The vegetation cover of the catchment is a particularly important influence on the health of a river. Vegetation stabilises the land and protects it from erosion by running water. It slows down runoff and takes up and evaporates water, thus reducing the frequency of flooding. Natural well-vegetated stream sides provide protective buffer strips, help filter runoff from adjacent land, reducing the amount of sediment, nutrients and toxic agricultural chemicals reaching the stream. They also help maintain stable channel banks and provide valuable habitats for many plants and animals.

In headwater streams, the bed of the channel may be dominated by boulders and rock outcrops, with deeper pools and shallow stretches where the water moves more quickly. In wooded valleys, streams accumulate dead wood or fallen trees, forming jams that give the channel a stepped profile.

Downstream, as the slope declines and the valley widens, the decreasing power of the river flow results in it depositing its sediment load. Bars of gravel form within the channel, sandy ridges – levees – along the channel banks, and spreads of finer sediments across the flood plain. Wide, shallow, braided channels form when a river transports a high sediment load, creating mid-channel bars which may become vegetated to form islands.

Low-energy, meandering rivers create areas of fast flow on the outside bank, whilst material is deposited at the inside bank where the water flows more slowly.

Natural channels with a high sediment load intertwine to form a braided pattern in the centre of a valley.

BOX 2.1 **Water in the UK**

Average annual rainfall varies from over 4000 mm on the mountains of Wales, the Lake District and western Scotland, to about 500 mm in parts of southern England. This is because the predominant weather pattern brings in rain from the Atlantic Ocean to the west. The amount of water lost by direct evaporation and transpiration from plant leaves is equivalent to between 400 mm and 500 mm of rainfall over most of the country, with a general northward decrease giving lowest values in northern Scotland. Higher values of 700–850 mm have been reported for forested catchments, in part reflecting high evaporation rates of water intercepted by the tree canopy. The balance of the rainfall less the evaporative losses is the effective precipitation, which both provides surface runoff and replenishes groundwater.

Across the UK, the wet uplands contrast with the dry lowlands, divided by an irregular line between the estuaries of the River Tees and River Exe. Mountains and uplands, underlain by ancient hard rocks that generally have low permeability, dominate the west where the highest summits reach 1300 m above sea level. In these areas, typically 75 per cent of effective precipitation becomes runoff and river flows respond quickly to rainfall. Lowland areas to the east are characterised by younger sedimentary rocks – sandstones, limestones and clays. Sandstones and limestones are generally permeable and often act as aquifers – rocks within which water moves sufficiently freely that it can be extracted via wells and boreholes. In these areas up to 75 per cent of effective precipitation may soak in, recharging the aquifers. Because less water is left to form surface runoff, river flows respond less dramatically to rainfall events in these areas.

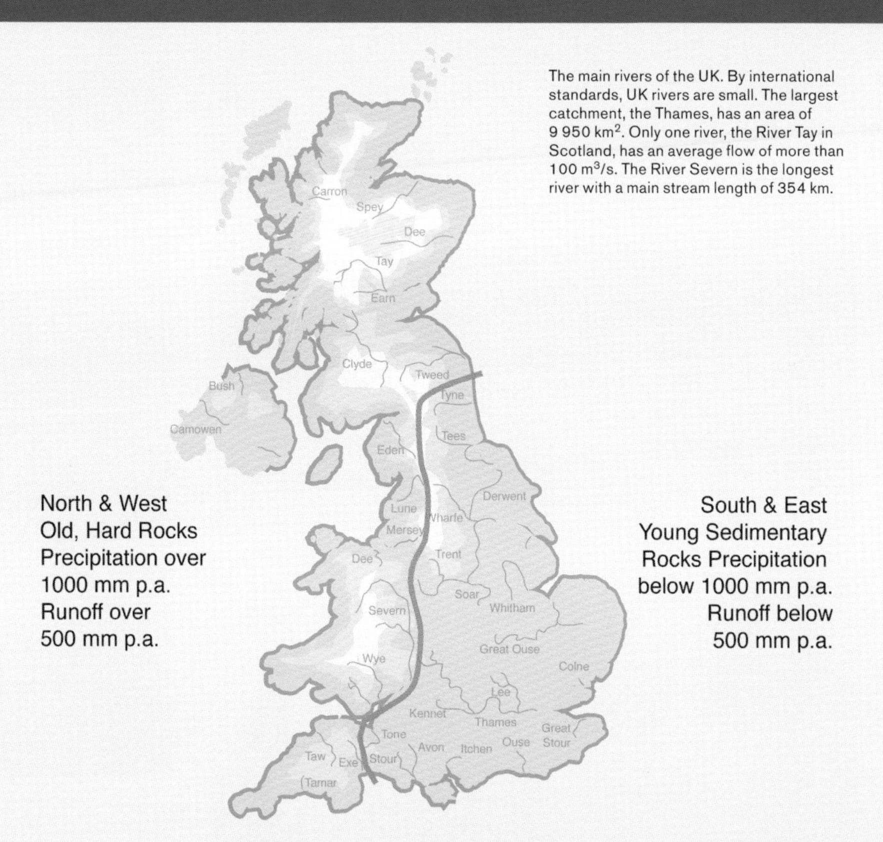

The main rivers of the UK. By international standards, UK rivers are small. The largest catchment, the Thames, has an area of 9 950 km^2. Only one river, the River Tay in Scotland, has an average flow of more than 100 m^3/s. The River Severn is the longest river with a main stream length of 354 km.

Graphs of river flow plotted against time are termed hydrographs. Different degrees of flow variability reflect the geological character of catchments. For example, in the Derbyshire Peak District, the hydrograph of the River Derwent, draining a catchment of relatively impermeable shales and hard sandstones, contrasts with that of the neighbouring River Wye, draining highly permeable Carboniferous limestone. The flow in response to a rainstorm in December 1965 peaked at over 150 m^3/s on the Derwent as a consequence of the importance of runoff in this catchment, but at only 38 m^3/s on the Wye. However, the minimum recorded flow on the Wye is more than double that on the Derwent, because of the importance of slowly moving groundwater in the Wye catchment.

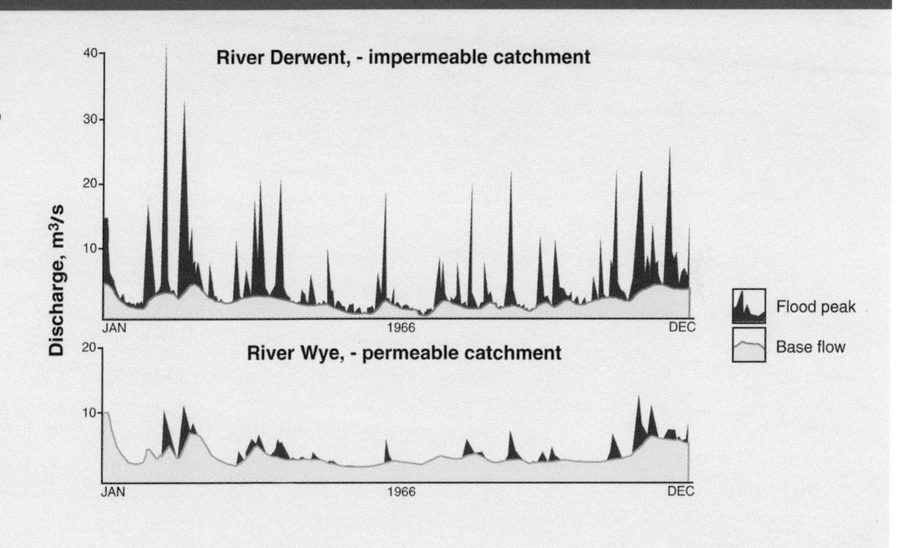

The flow of the River Derwent has many more flood peaks than does the flow of the adjacent River Wye, because of the largely impermeable character of the River Derwent catchment compared with the permeable River Wye basin.

BOX 2.2 **Energy, oxygen and food chains**

Within an ecosystem, plants and animals are linked in a food chain, where each is the food of the next member of the chain. The energy from the sun is used by plants (known as primary producers), which are then eaten by herbivores and primary consumers or which die and contribute to a pool of detritus (leaves, twigs and even whole trees) that slowly decomposes. Organic matter entering streams is washed in from the hill slopes and channel margins to supplement the primary production by aquatic algae and plants.

In running-water ecosystems, bacteria and fungi break down organic matter, such as leaf litter in mountain streams and submerged plants from shallow downstream waters. This provides food for organisms that dwell on the river bottom, like insects and snails. These animals are then eaten by a variety of carnivorous animals (the secondary consumers) including invertebrates, fish, amphibians, birds and mammals.

In the process of decomposing organic matter, bacteria remove oxygen from the water, creating a biochemical oxygen demand (BOD). Oxygen used in decomposition is replaced by that gained by exchange at the water surface, especially over weirs and in fast, turbulent stretches. Oxygen is also gained from plants and algae by photosynthesis during daylight hours. Thus, natural rivers can be described as "self-purifying".

Further downstream, as the river gradient declines, the channel tends to wander across a wide flood plain in a series of meanders. These flood plains form natural storage for flood water at times when runoff exceeds the capacity of the river channel. Across the flood plain, deep water in abandoned channels – oxbow lakes – and wetlands formed behind levees all contribute to a rich mosaic of habitats for wildlife.

Living rivers

As water flows increase from headwaters to mouth, the character of a river changes. Temperatures increase with falling altitude and reduced shading as channels widen; the water becomes richer in nutrients as the catchment becomes broader. The shape of the channel also changes with changing flow, sediment load and valley gradient. The distributions of animals and plants show typical patterns along streams and rivers that reflect these downstream changes. Each of the habitats will be occupied by animal and plant communities that have adapted to the conditions. Any change in these communities may indicate a decline in the health of the watercourse. Many animals and plants depend upon particular habitats for all or part of their life cycles. By studying protected areas and the remaining semi-natural, rural catchments, it is possible to reconstruct what the different streams and rivers of the UK must have been like before industrialisation and urbanisation.

The springs of groundwater-dominated streams can have considerable ecological interest, but many have been lost or severely degraded by urban development. Others are highly vulnerable to future groundwater abstraction. The Babingley Brook on the chalk aquifer of north Norfolk is one example of a healthy springhead. It has a variety of habitats with temporary and permanent ponds, reedbeds, wet meadow and wet woodland. A diverse invertebrate fauna includes over 100 species of beetles and at least 55 species of fly dependent upon wet or damp habitats. The area is managed to conserve its value for water vole, breeding wetland birds, migrant wetland birds including significant numbers of geese and duck, amphibians, and at least 40 species of plants including common spotted orchids and early marsh orchids.

The headwaters of upland rivers, shaded by bankside oak, ash and alder trees, have fast cool flows and well-oxygenated, silt-free waters that provide suitable habitat for trout and spawning grounds for salmon. This is the home of the dipper, a bird that has become totally dependent on the stream, doing almost all its feeding on aquatic insects below the surface. Another common bird of upland streams, the grey wagtail, feeds mainly on adult insects emerging from the water.

Along the wandering middle river, fish such as grayling and minnow characterise the faster reaches. Open grassy areas of gravel bars provide summer habitat for nesting

Water voles were once found near watercourses throughout Britain, but today they are a vanishing species. Loss of habitat and pollution are thought to have contributed to this decline. (Hugh Clark)

The northern marsh orchid is a plant of bogs and wet grasslands from the Midlands northwards. (Iain Wright)

Unable to survive in polluted waters, the salmon has become the symbol of a healthy river.

The dipper, commonly found feeding on aquatic insects in upland rivers. (www.rspb-images.com, Mark Hamblin)

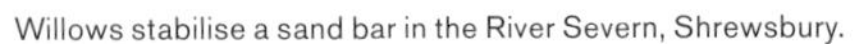

Willows stabilise a sand bar in the River Severn, Shrewsbury.

Snake's head fritillary, a member of the lily family, favours wet meadows.
It is now a rare plant in the wild. (Peter Stiles Photography)

The redshank, a wading bird, which feeds on mud banks of lowland rivers. (www.rspb-images.com, Bob Glover)

lapwings and common sandpipers. The areas of exposed sediments provide an important habitat for a wide range of flies, beetles and spiders.

The sluggish lower river, with its network of backwaters, provides a wide range of habitats for bream, dace, perch, pike, tench and carp. Seasonal floods connect the river channel to a variety of wetland and still-water habitats across the flood plain, providing nursery areas for fish and feeding grounds for migratory birds. Flood plains also provide habitats for wetland plants and animals that depend upon regular flooding to bring nutrients or to protect them from competition with dry-land species.

In winter, shallow, flooded washlands attract a wide variety of wildfowl, drawn by rich grazing and secure roosting sites. The water meadows on the great fenland rivers Ouse and Nene still attract vast numbers of winter visitors: more than 30 000 wigeon, 3000 Bewick's swans, 4000 teal and many other species. The summer water meadows are also full of breeding birds such as the common snipe. The banks of oxbow lakes provide favoured habitat for reed buntings, attracted by the abundant cover for nesting and the prolific crop of seeds and fruits. Under low water levels in summer, exposed patches of mud attract waders, such as ruff and spotted redshank, returning from their northern breeding grounds.

At the end of the river, the estuary receives all the runoff from the catchment and the water is rich in nutrients. This resource is the food base for millions of small organisms (worms, snails and shrimps) in every square metre of mud. In turn, these organisms provide a rich source of food for fish and birds. Almost 90 species of fish have been recorded from the upper parts of the Severn estuary. Oystercatchers are characteristic of the estuarine tideline and are specialists in exploiting shellfish for food. Like flood plains, estuarine marshes offer rich habitat for flocks of wildfowl and waders escaping the harsh winter conditions of the Arctic. The grasslands and marshes of the River Severn, for example, are particularly attractive to overwintering geese and swans from as far afield as Siberia.

Oystercatchers are mainly a shoreline bird, found on sandbanks and mudflats in search of shellfish, but they breed locally by rivers and gravel pits. (www.rspb-images.com, Chris Gomersall)

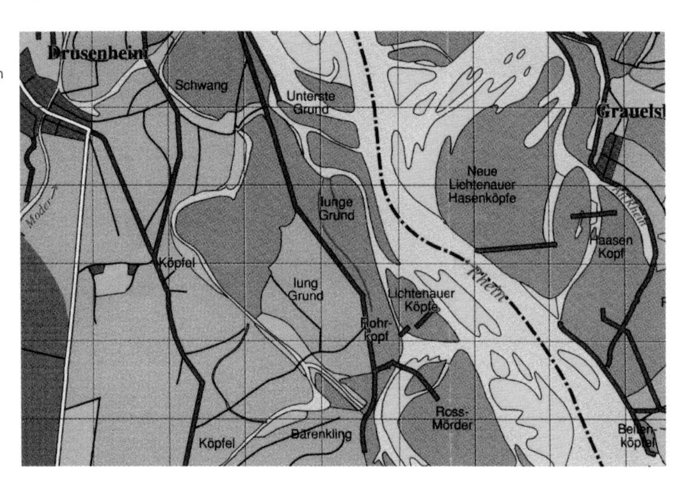

The River Rhine, Germany, in 1838 and 1989. Channel engineering has made the modern river narrower and deeper than the natural channel, permitting waterside industrialisation. (WWF-Auen-Institut)

From prehistory to the modern day

The history of rivers reflects both changing climate and human activity. Over the past 100 000 years, the Ice Ages have had a major impact. North of the Thames basin, retreating glaciers left thick sheets of boulder clay, sands and gravels. Rivers cut their valleys down to sea levels that were more than 100 m lower than today. After the retreat of the glaciers, about 14 000 years ago, wildwoods of lime, elm, holly, ash, beech and oak came to dominate the landscape of the UK. Then, about 6 000 years ago, the arrival of Neolithic peoples heralded the conversion of the natural wildwood landscape to open countryside dominated by farmland. By Roman times, much of the lowlands had experienced a long history of farming while moorland now dominated the uplands. For nearly 2000 years, the rivers of the UK were in balance with the low-intensity agricultural landscape.

Progressively, problems of over-fishing, pollution and navigation had local but notable impacts. In the UK, evidence of over-fishing of salmon dates from the 13th century; small urban streams such as the Walbrook in London were reported to be suffering from pollution by the end of the 14th century; and river works for navigation were having a significant effect from the mid-17th century.

Once the Industrial Revolution came (in the early 19th century in the UK) the rapid increase in population led to a huge increase in the demand for water to supply industry and human needs. Demands for food altered agricultural practices,

BOX 2.3 **Past changes of climate**

The world's climate has never been static. At the end of the last Ice Age, about 14 000 years ago, global temperatures were about 4°C below the mid-20th century average temperature. At their maximum, in about 4000 BC, average temperatures were more than 1.5°C above this. A cold phase followed from about 1500 to 500 BC, then the Medieval Warm Period from 1050 to 1350 AD, and then the Little Ice Age from about 1500 to 1850 AD, when global temperatures fell by up to 0.5°C below the mid-20th century average. Climate is driven by variations in the amount of solar radiation received by the Earth's atmosphere.

Irregularities in the Earth's orbit around the sun explain some of the variations. These cause changes with distinct periodicities: 100 000 years as a result of the eccentricity of the Earth's orbit around the sun; 41 000 years because of the Earth's axial tilt or obliquity; and 21 000 years caused by the precession of the equinoxes. The climate is also sensitive to the composition of the atmosphere, including greenhouse gases such as carbon dioxide and methane as well as dust concentrations, and some authors have associated major changes of climate – and the extinction of the dinosaurs – with massive volcanic eruptions and meteorite impacts.

During the 17th and 18th centuries, frost fairs on the frozen Thames were a common occurrence.

and demands for space by the urban dweller affected rural landscapes as never before. From this time until today, rivers in the UK became dominated by human influences.

THE CATCHMENT TODAY

Most modern-day catchments bear little resemblance to natural catchments. The demands of urban populations have induced continuing changes in rural landscapes. Intensively cultivated crops and managed grasslands dominate lowland catchments. Coniferous plantation forests cover upland hillsides, and valleys have been impounded by dams. Buildings and impermeable tarmac cover large areas of lowland, and docklands have replaced estuarine marshes. All these activities within catchments influence the character of a river as it flows through a town or city.

BOX 2.4 **Groundwater depletion and rebound**

For urban areas situated over aquifers, groundwater abstraction met the needs of urban and industrial expansion in the late 19th and early 20th centuries. The rate of abstraction often exceeded the rate of natural replenishment, however, resulting in a significant drop in the water table level. Examples can be found in Birmingham, Liverpool and London. Around London concerns for the decline in spring flows from the important chalk aquifer were raised as early as 1850. Below Trafalgar Square, as groundwater levels fell, the springs and water-supply boreholes ran dry. The drilling of deeper boreholes was expensive and many were contaminated by salt water intruding into the aquifer from the River Thames. The artificially low water table encouraged the building of deep structures such as basements and tunnels. However, continued growth and inevitable pollution of urban groundwater made municipalities seek more easily secured surface-water supplies. This effect, combined with the downturn in the UK manufacturing base, meant groundwater abstraction declined from its peak usage in the immediate post-war period and water tables began to rise. This has resulted in potential flooding problems for buildings and infrastructure built below ground level, as well as subsidence problems in places where soil properties have altered. In addition, pollutants may be flushed from the soil and ground as groundwater discharges once more into rivers.

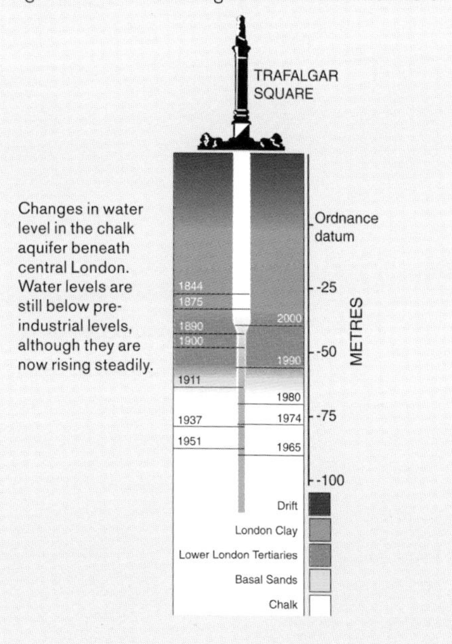

Changes in water level in the chalk aquifer beneath central London. Water levels are still below pre-industrial levels, although they are now rising steadily.

Demands for water

As urban communities grew, so did their need for water. This growing demand was met by increasing abstractions from rivers and then by river regulation using dams and reservoirs.

Groundwater, exploited by wells and boreholes, is an important resource throughout the central, southern and eastern regions of the UK, and abstractions from these groundwaters have caused low flows in rivers and the drying up of springs. Replenishment of these groundwater supplies is often prevented by traditional urban drainage – the covering of the ground with impermeable tarmac and concrete and the diversion of runoff into the stormwater drainage system, leave little opportunity for rain to enter the ground.

From the late 18th century, towns at the centre of the Industrial Revolution, including Sheffield, Manchester, Bolton and Halifax, and elsewhere, such as Edinburgh, began to exploit small streams in upland areas in preference to their increasingly dirty local rivers. Streams were impounded by dams, creating reservoirs to store winter runoff, to augment supplies in summer and to supply a continuous flow of clean water to the urban area by gravity. In several cases water was transferred to a town outside the catchment, causing a severe reduction in the flows in the supplying river and swelling the receiving river.

Today, most major upland rivers in the UK are controlled by dams. Even "lowland" rivers such as the Trent, Great Ouse and Medway are heavily regulated. Dams and reservoirs

Over-abstraction to meet the increased water needs of expanding urban communities can result in dry rivers (River Misbourne at Chalfont St Giles).

Dams, such as the Rheidol dam in Wales, provide a reliable source of water for domestic use and also industrial purposes such as hydroelectric power generation.

introduce barriers that disrupt the natural river system. Most obviously they obstruct the passage of migratory fish such as salmon. However, they also reduce or eliminate regular flooding, essential for transporting sediment along the river and maintaining the dynamic channel form and the mosaic of physical habitats this provides. River temperatures are changed, and the type of organic matter that drives the food chain is altered. The reservoir itself introduces a new type of habitat, but unlike natural lakes, the water levels of many reservoirs have wide variations, which makes the shores inhospitable for many organisms.

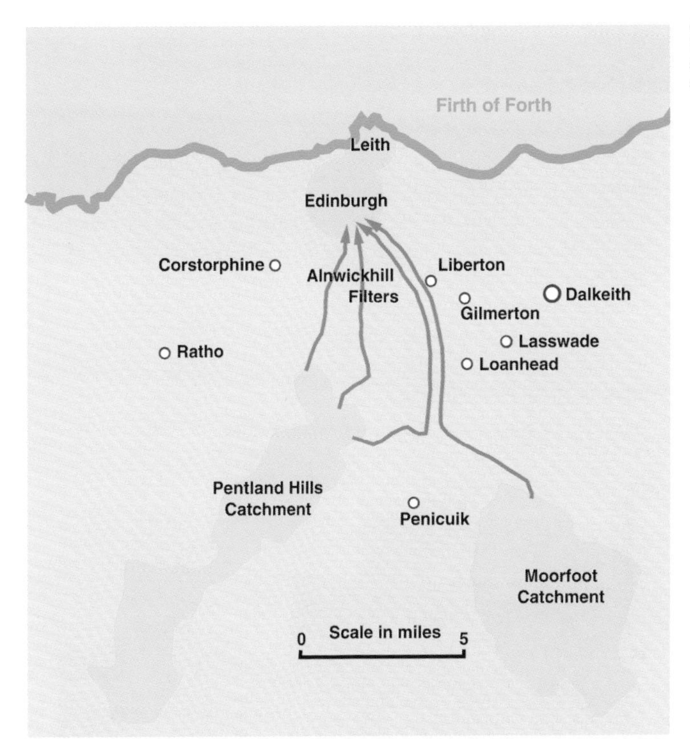

Catchwaters, such as those in Edinburgh, draw water from the surrounding upland areas to supply the urban population.

Floods

A natural river can be expected to overflow its banks. However, extreme floods are rare events, although it is possible for several severe floods to occur in a short period of time. Changes in the use of land, such as the removal of

BOX 2.5 **Floods return**

The likelihood of flood inundation is described by the flood return period. The flood that is likely to be equalled or exceeded on average once in a hundred years is termed the 100-year flood. However, a flood having a return period of once in 100 years does not imply any regularity of recurrence. Two 100-year floods could occur in two consecutive years, or even in the same year. Where the return period is long, the flood is likely to be large, but less likely to happen in any one year. No matter how high the maximum recorded flood, an even higher flood may occur sometime in the future. During the floods of autumn 2000, several locations were flooded for the third or fourth time in two years, and York suffered its highest flood level since the 17th century.

Shrewsbury Abbey was flooded by the River Severn in November 2000, the worst floods since 1947 (see also page 61).

Soil erosion, caused by surface runoff, can be accelerated by poor land management practices including intensive agriculture.

native woodland, drainage and intensive cultivation of farmland, and especially urbanisation, have increased rates of flood runoff in many catchments. The result of this is that higher peak flows occur more often than in an undeveloped catchment.

Pollution from fields

One consequence of increased runoff is soil erosion, especially from intensively cultivated areas. Soil is not only essential for food production and an important habitat for many animals and plants, but it also has important functions within the catchment itself. The soil can filter, transform and neutralise potentially polluting substances, including acid rain; and it can act as an important reservoir for water. Soil erosion is a natural process, but some land management practices can accelerate erosion rates to abnormally high

BOX 2.6 **Problems with algae**

The most visible of the effects of human activity on surface waters is in the formation of "algal blooms", promoted by the growth of micro-organisms such as bacteria, algae and diatoms. In slow-flowing or standing waters, floating (planktonic) algae can turn the water bright green. Flowing waters can become choked with "blanket weeds", such as the alga *Cladophora*.

Plants and animals need dissolved oxygen to live. At night, plants consume oxygen from the water and massive plant growth can cause severe diurnal cycles of oxygen depletion. When large mats of algae die, rapid bacterial decomposition can severely deplete oxygen levels, killing invertebrates and fish.

As well as looking unsightly, algae interfere with recreational uses such as angling and water sports, and impede flood flows. Some of the algae produce potent toxins, which can pose a hazard to those practising water contact sports and to animals that drink the water. Where water is abstracted for public supply, it has to be treated to remove the algae and the toxins.

Problems with algae are primarily caused by the process of eutrophication - the enrichment of waters with nutrients, especially phosphorus and nitrogen. Today, sources of phosphorus are dominated by human and household wastes (43 per cent), with detergents contributing 19 per cent. Another 43 per cent comes from agriculture (livestock 29 per cent, fertilisers 14 per cent). Nitrogen inputs are also dominated by agriculture and urban waste water.

Today, the most elevated levels of phosphorus (as phosphate) and nitrogen (as nitrate and ammonia) are found in the rivers of the Midlands, East Anglia and south east England, and in Northern Ireland, that is, in highly populated and intensively farmed areas. By 1999 the UK Government had designated 2540 km of 62 rivers and canals in England and Wales – mainly in eastern England – as Sensitive Areas (Eutrophic) under the EC Urban Waste Water Treatment Directive. As a result, phosphorus reduction treatment will be required at around 150 large sewage treatment plants to reduce nutrient inputs to these waters.

A eutrophic river turned green by algal bloom resulting from an excess of inorganic nutrients, such as nitrogen and phosphorus in the water. This process can occur naturally but is often the result of human activities, such as fertiliser runoff and sewage discharge.

levels. This soil then accumulates in rivers, damaging the riverbed habitat and eliminating the organisms that live there, on which many others depend. This can result in the river containing very few living organisms.

Harvesting, soil disturbance, drainage and fertiliser applications can all influence nutrient levels in streams. Phosphorus and nitrogen are essential plant nutrients that are supplied to the soil by manures and fertilisers in order to maintain crop yields. In natural aquatic systems, upland rivers contain very low levels of nutrients, and the levels increase downstream. Plants such as algae are always present but their growth is limited by the availability of nutrients. Human activities can greatly increase nutrient levels, which can cause enormous growths of algae and higher water plants. In severe cases, algae can cover the riverbed, destroying habitat for most of the organisms that live there.

Coniferous forests can increase the acidification of streams by transferring atmospheric pollutants to the ground.

BOX 2.7 **Acid water**

A common measure of acidity is pH. A decrease in the pH value indicates an increase in the acidity of water. Water is neutral at pH 7 but rainfall is naturally acidic, because of dissolved carbon dioxide which produces a weak solution of carbonic acid. Under natural conditions, the pH of rainfall varies between 5 and 7. pH values below 5.6 often indicate atmospheric pollution; "acid rain" can have a pH of less than 4.5. As precipitation becomes runoff the water reacts with the soils and rocks of the catchment. For natural waters, pH values commonly range between pH 5 and pH 9, although some hot springs can be as low as pH 2. Streams draining lime-rich soils and rocks, such as chalk and limestone, are usually neutral or slightly alkaline, pH 7 to 8. Waters affected by acidification can have a pH value below 4.5.

Acid attack

The burning of fuels, especially coal and oil products, and some industrial processes release acid gases into the atmosphere. These gases dissolve in rain, which can contain up to five times as much acid as it would in a pristine environment. In some upland catchments the soils lack capacity to neutralise the acid rain, which can leach natural aluminium from the soil, and this can be acutely toxic to many forms of aquatic life. The expansion of coniferous plantations has exacerbated the effects of acid rain. By 1995 some 12 000 km of upland streams in Wales were thought to have been degraded by high acidity. In Scotland, acidification is the third most significant cause of river pollution, accounting for 12 per cent of polluted river length. As a consequence of this, waters with once productive fisheries have become almost devoid of fish. Measures have now been taken to reduce acid emission to the atmosphere, including low-sulphur diesel and catalytic converters on cars, but recovery of soils, rivers and streams affected by acidification may take many years.

Otters inhabit rivers with clean water, healthy fish stocks and good vegetation. Their presence on a riverbank is a good indication of freshwater quality.

BOX 2.8 **Threatened species**

Across Europe, 17 of 222 recorded river fish species are globally endangered or vulnerable. Migratory species such as salmon and sturgeon are directly affected by dams which block their journeys, pollution, and loss of spawning and nursery grounds as a result of sediment deposition. The geographic range of the otter has been drastically reduced this century by habitat destruction, pollution and hunting.

In the UK, between 10 and 20 per cent of native species are considered threatened. Urban development, transport development, forestry expansion and intensification of agricultural production are principal causes. In rivers and streams, habitat loss and pollution have had significant effects on biodiversity. For example, just over half of the dragonfly species declined in geographical distribution between the early 1970s and late 1980s. Dragonflies rely on water – rivers, streams, ditches, lakes and ponds, bogs, heaths and moorland – and some are highly specialised to a single habitat.

Damage to wildlife

In natural rivers and streams, communities of animals and plants comprise a large number of different species with relatively few individuals of each species. Poor water or loss of habitat reduces the diversity of animals and plants present. As conditions deteriorate, chain reactions occur because the plants and animals living in an area are often dependent upon each other, for example, for food. An unhealthy community is characterised by a restricted number of species that can tolerate poor conditions, though there may be a large number of individuals of each of these species.

URBAN RIVERS

The Industrial Revolution in the UK in the early 19th century led to a spectacular rise in urban population (defined as those living in towns of more than 5000 people) from approximately 3.1 million in 1800 to almost 28 million a century later. This represented both an unprecedented increase in native populations and a large-scale migration into towns and cities. In London alone, the population almost trebled to 2.4 million in the 50 years up to 1851, but provincial industrial towns around the tidal estuaries and inland coalfields also became major urban centres.

The inevitable result of such rapid, unplanned expansion was insanitary and unhealthy housing in the cities and the emergence of slums. Open ditches and streams had long been

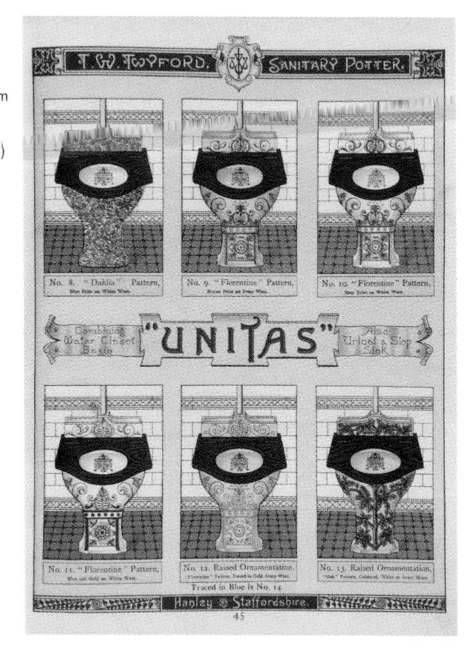

Images of the UNITAS closet taken from T.W. Twyford's 1888 catalogue. The UNITAS, designed and produced by Thomas William Twyford in 1883, was the first all-ceramic one-piece pedestal wash out (flushing) closet and the forerunner of modern-day toilets. The introduction of flushing toilets had a significant impact on the amount of sewage discharged into rivers and streams. (Reproduced with kind permission of Twyford Bathrooms)

The concrete lined channel, choked with rubbish and green with algae, typified the degeneration of urban rivers when the photograph was taken, in the early 1990s. (The Lower Lea Project)

receptacles for domestic sewage and trade refuse, as well as rubbish from street surfaces, and their flowing water provided a simple means of waste disposal. With growing populations the ability of streams to disperse, dilute and assimilate organic wastes became overwhelmed, and channels often became blocked. The expense of cleaning and maintenance led to widespread neglect. Plague and cholera were the inevitable consequences.

The cesspit was the traditional method for disposing of domestic sewage. The accumulated sewage was removed periodically for eventual use on market gardens and farms around the city. In poorer areas, cesspits went unemptied for years on end. Seepage and overflows to streams were common but the pollution was spread along the length of the streams and some measure of self-cleaning took place. Ironically, the widespread "improvement" in sewage management with the use of the water closet from the 1820s and the construction of sewer systems had dramatic impacts on inland waterways, creating large point sources of sewage discharge directly into streams and rivers. Rivers could not cope with this heavy load and the effect on the ecology downstream was catastrophic.

"Good practice" in urban drainage during the second half of the 18th century advocated the culverting of all streams and ditches. The growing pressure on urban space and the increasing use of road transport added to demands to cover over the open sewers, both for health reasons and to improve the urban transport network. Thus, many streams became

BOX 2.9 **Sewage decomposition**

Faecal and organic pollution are the oldest forms of pollution to affect rivers. Organic matter, and especially sewage, is a particular problem for urban rivers. Bacteria in the water will decompose sewage, using up oxygen in the process. A normal end-product of the decomposition of animal protein, faecal matter and urine is ammonia, which is toxic to fish and other organisms at high concentrations. Bacteria eventually convert ammonia to nitrate, using more oxygen in the process, and the nitrate contributes to eutrophication. When large amounts of organic matter are involved, so much oxygen may be used by the bacteria that the water may be deprived of oxygen. Under such anoxic conditions, a deposit of highly organic mud can form on the bed. In the absence of free oxygen, bacteria will use other oxygen sources, such as carbohydrate, sulphate and nitrate in the mud, to produce methane, hydrogen sulphide and nitrogen gas. Hydrogen sulphide has a pungent bad-egg smell and turns water black due to the precipitation of insoluble iron sulphide. Methane and nitrogen gas can cause "rising sludge", the noxious bottom mud that floats up to the surface of the water.

underground sewers; their courses became roads and railways. Those streams that survived into the mid- 20th century were straight-jacketed in concrete-lined channels edged by embankments to accelerate their flows – and their capacity to transport rubbish – downstream from the urban area.

An early picture of the River Thames, showing shipping at the confluence with the Fleet. (Reproduced with kind permission of Thames Water)

The fall and rise of London's river

The River Thames has a long history of pollution, but until the late 18th century the large flow in the main river was adequate to dilute and disperse the wastes from towns in the catchment. By 1771 when Tobias Smollett wrote to Humphry Clinker, the river was in rapid decline:

> ... the river Thames, impregnated by all the filth of London and Westminster. Human excrement is the least offensive part of the concrete which is composed of all the drugs, minerals, and poisons, used in mechanics and manufacture, enriched with the putrefying carcasses of beasts and men; and mixed with the scourings of all wash-tubs, kennels, and common sewers.

BOX 2.10 **Death of the Fleet**

The River Fleet in central London is the largest and most important of the Thames's lost tributaries. Rising on Hampstead Heath and in Highgate, the river flows past Camden Town, King's Cross and Clerkenwell to join the Thames near Blackfriars Bridge. Until the end of the 18th century the river provided an abundant water supply, power to drive a series of water mills, an important navigation route, and a line of defence to protect the City of London's northern approaches. Its flow was derived from a series of wells. From 1760, Bagnigge Wells (now about 61-63 King's Cross Road) were a popular country resort for pleasure-seeking Londoners attracted by the medicinal waters and tea drinking.

The River Fleet, which flows from Hampstead Ponds to Kings Cross, shown here at St Pancras, before the expansion of London covered the river. (Reproduced with kind permission of Thames Water)

Downstream, within the city, the river had a very different reputation. From the 13th century the Fleet was an important industrial corridor and a haven for crime, characterised by poverty and squalor. The channel became an open sewer.

In medieval London wherever there was running water, latrines were built over it as the easiest method of disposal. In Ben Johnson's *On the Famous Voyage* published in 1616, he describes the foulness of a voyage up the Fleet: "When each privies seate is fill'd with buttock? And the walls doe sweate urine and plaisters? . . . When their oares did once stirre, belch'd forth an ayre as hot as at the muster of all your night-tubs." The river was not only unsightly and unbearably smelly, but also became a source of epidemic disease. The plague of 1665 hit every parish along the lower river.

By 1733 the Fleet navigation had become obsolete, and choked with silt and rubbish. The wharves became roads and the canal was covered over. In 1846 the polluted river "– quite literally – blew up, its rancid and foetid gases bursting out into the street above.... in Clerkenwell three poorhouses were swept away in a tidal wave of sewage" (Trench and Hillman, 1984). Finally, the river was lost beneath the expanding urban infrastructure. The Fleet had been turned into a large underground sewer measuring 5.6 m by 3.7 m at its mouth, and carrying an estimated 8 million cubic metres of sewage each year.

The Fleet ditch (1844) behind the Red Lion, reputed to have been the hideout of Dick Turpin. (Reproduced with kind permission of Thames Water)

Spoonbills were once common in the marshes and reedbeds of the River Thames. (www.rspb-images.com, Gordon Langsbury)

A major problem was the waste from the 150 slaughterhouses, as well as human excrement. Worse was to come. Street gas-lighting, introduced in 1807, used coalgas produced at a number of small works sited close to the river into which toxic waste products (ammonia, phenol, tar) were discharged.

Before the Industrial Revolution and the growth of London, the inner estuary below London Bridge was a wilderness of marshes and reedbeds with diverse wildlife. Bitterns, bearded reedlings, spoonbills and montagu's harrier would have been common. Salmon, smelt, shad, flounder, eel and whitebait, together with cultured mussels and oysters, formed the basis of important fisheries.

In his survey of fish in the River Thames in 1859, William Yarrell wrote that the smelt had become uncommon and records: "The last Thames Salmon I have note of was taken in June 1833." The virtual destruction of the natural ecology of the River Thames between 1800 and 1850 was the result of attempts to improve public health and to provide more land for the expanding urban population.

Reclamation of the marshes

Low-lying marshes along the river were ideal for the construction of docks. Between 1810 and 1886 new docks were completed and land was also lost to industry and housing. The bird life of the upper tidal river was reduced to a few mallard and mute swans feeding on spillages from the grain wharves, and a variety of gulls scavenging in the sewer-like conditions.

Improved sanitation

By 1855 the population of the growing urban area had passed 2 million. In response to cholera epidemics between 1831 and 1866, which killed more than 37 000 people, the Metropolitan Commission of Sewers was formed in 1847 to abolish cesspits, clear the sewers and improve house drainage. Water closets were now popular with the middle classes, and one commissioner, Edwin Chadwick, a great social reformer, advanced the regular flushing of sewers with water. With ever-increasing efficiency, sewage was flushed into an increasingly overloaded sewer system and discharged directly to the Thames.

The problem was exacerbated by the way sewers discharged to the river, at the time of low tide. As the tide rose, sewage was held back in the sewers, causing it to stagnate for 18 out of every 24 hours. Sewage discharged at low tide was carried upstream by the rising tide, and brought back to London by the following ebb tide to mix with the next day's discharges. Especially during summer, low freshwater flows were incapable of diluting and dispersing the accumulating volume of wastes, and vast quantities of foul and offensive mud were exposed along the banks of the river. Although this system brought about a great improvement in human health, the ecological damage to the river was severe.

The solution: transfer the problem downstream

The political will to finance major new schemes was slow to be realised because there was a lack of scientific proof of the link between waste water and public health. And the problem had not reached crisis proportions. It took the hot summer of 1858 – the year of the Great Stink – to persuade politicians of the need for action. The Times of 3rd July 1858 reported that the Thames stank so badly that Parliament was nearly closed and members of a committee led by the Chancellor of the Exchequer were observed to rush from the "pestilential odour".

The engineer Sir Joseph Bazalgette designed and laid out new works. Three lines of sewers were laid to intercept the existing sewers, and convey their sewage to outfalls into the river 17 km below London Bridge. The principle behind Bazalgette's scheme was simple: gravity. Pumping stations at various points were used to raise waters from the lower-level sewers to join the higher-level northern and southern interceptor sewers leading to the two major sewage treatment works at Becton and Crossness. By 1865 there were about 2000 km of sewers in London and 132 km of intercepting sewers. The main drainage works of London, which had cost over £4 million, were cited at the time as "a magnificent instance of successful drainage".

The state of the River Thames attracted widespread criticism, such as this cartoon drawn by Tenniel for Punch in 1858.

The river dies

The Bazalgette system quickly became overloaded and, despite the introduction of new plant, treatment was inadequate to meet the demands of a rapidly expanding city. Damage to the sewers and treatment works during the two world wars saw the Thames deteriorate further, to the extent that the water contained no dissolved oxygen in summer. This killed virtually all wildlife in the river.

The decline in the river continued until the mid-20th century, when the whole section of the Thames through London was virtually lifeless. A survey of remnant marshes between 1949 and 1950 found few wildfowl upstream of the sewer outfalls. The birdless state of the upper tidal Thames continued through the mid-1960s. The survey of fish in the London area by Alwyne Wheeler in 1957 showed that there was no established fish population between Kew Bridge and at least as far downstream as Gravesend, a distance of some 50 km.

Studies in the early 1960s showed that the Bazalgette scheme actually exacerbated the pollution problem by concentrating inadequately treated sewage which, because of the enclosed tidal nature of the river, became trapped in a reach only about 5 km below the discharge point.

Towards recovery

New scientific understanding of the tidal flow of the river in the London area showed that the treatment of sewage had to be improved. New processes were added, including advanced filtration and aeration to improve the quality of the waste water discharges. The solid matter was also treated to render it non-toxic before dumping at sea (sea dumping has now ceased). Many small, inefficient works were closed, and the flows redirected through larger, modern regional works. By 1970 the 190 small ineffective sewage works that had existed within the London area in 1935 were reduced to 12 improved regional works. By 1975 minimum oxygen levels during summer, which in 1963 were close to being anaerobic for 18 km below London Bridge, had recovered. Within 10 years fish and birds were returning to the tidal Thames.

Regeneration of the River Thames is indicated by the return of wildlife such as flocks of pochard. (www.rspb-images.com, George McCarthy)

The 40 km reach below London Bridge benefited most. Between 1967 and 1973 a total of 72 species of fish were identified in the river. Six migratory species, including smelt, penetrated upstream to at least 10 km above London Bridge. In 1974 a few salmon were reported to have returned to the lower tidal river. Signs of change in the bird population along the upper tidal river were noted during the mild winter of 1968 with the arrival of two large flocks of pochard with numbers of shelduck, pintail and tufted duck. Once again the upper tidal river became a refuge for many species of water birds, including a wintering population of up to 10 000 wildfowl and 12 000 waders.

While the river was barren and stinking, the waterfront had little attraction for development. In cleaning the river and restoring the ecology, the foundation for regenerating the waterfront had been established at last.

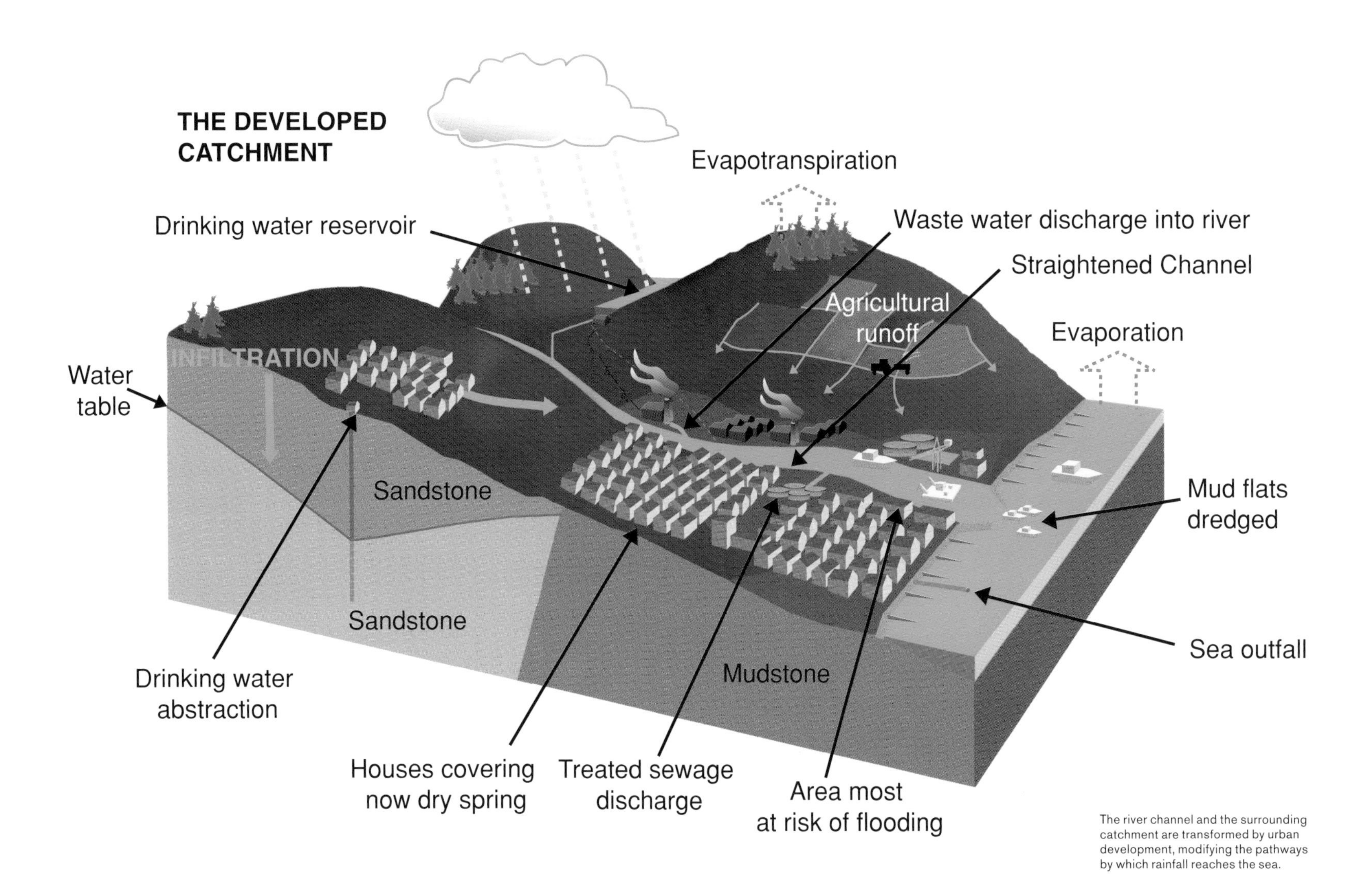

The river channel and the surrounding catchment are transformed by urban development, modifying the pathways by which rainfall reaches the sea.

3 Urban Areas Today: Problems And Solutions

A better understanding of urban water problems has led to a new vision for urban rivers. Once perceived as symbols of decay and pollution and a danger to public health, urban rivers should now be seen as one catalyst for urban regeneration. However, public concern remains. In England, Environment Agency consultation with groups and individuals concerned about the rivers and streams in their local areas identifies three key issues:

Pollution wrong connections, stormwater, pollution incidents, eutrophication, runoff from contaminated land and persistent chemicals;

Flows flooding, rising groundwater and low flows;

Amenity and ecology litter and illegal tipping, derelict land, algae, poor fishing, degraded habitats and drain-like channels, conservation of rare species and the invasion of non-native species.

This section explores the reasons why these issues are of concern and the options available for their management. All the issues relate in some way to the urban water cycle. Urban areas use huge volumes of water for domestic and industrial use, and create large volumes of waste water for discharge to rivers and streams. In addition, the extensive urban surfaces and drainage and sewer systems produce rapid runoff during rainfalls. Thus, the urban water cycle is characterised by pollution and flooding.

Box 3.1 **Reducing domestic water use**

In the past 30 years the amount of water used for domestic purposes has increased by 65 per cent, but there are many simple steps we can take to reduce the amount of water we use. For example, some showers use only a third of the water required for a bath and using a glass of water to rinse teeth after brushing, rather than leaving the tap running, could save as much as 9000 litres per person per year! A modern toilet uses 7 litres per flush or less, compared with older ones using 9 litres or more. Reducing the volume of water in the cistern, by fitting a "Hippo" bag or plastic drink bottle, would further reduce the amount of water used by each flush.

The two biggest water users in the kitchen are the automatic washing machine and dishwasher. Modern washing machines use less than 60 litres per cycle, compared with over 100 litres per cycle for older machines; by using these appliances only for a full load, it is possible to reduce the amount of water, power and detergent used.

Britain is one of very few developed and developing countries that does not meter all mains water consumption. More widespread use of water meters, increasingly installed as new houses are built, will encourage more economical use of water. A reduction of domestic use to 100 litres per person per day, which is possible, would leave an extra 40000 litres/second in rivers. It would also reduce the volume of waste water and the associated cost of treatment.

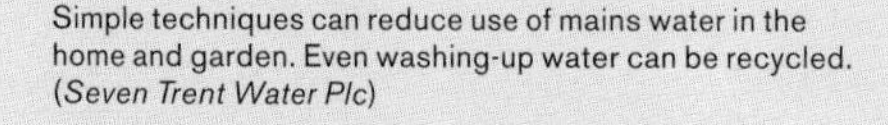

Simple techniques can reduce use of mains water in the home and garden. Even washing-up water can be recycled. (*Seven Trent Water Plc*)

Box 3.2 **Collecting rain**

Rainfall collection systems and the storage of filtered rainwater in tanks beneath houses can supply many purposes apart from cooking and drinking. Such schemes can provide essential storage of water in winter, when excess roof drainage can cause flooding, for supply during the summer when water is a valuable resource.

The Millennium Village on the Greenwich Peninsula aims to minimise water use. Its 1377 homes have water-saving devices such as efficient showers and toilets. The Dome itself was designed to collect the millions of gallons of rainwater that fall on it every year for use in the 650 visitor toilets and washrooms. The distinctive roof of the Sainsbury's superstore has also been designed to collect water, which is treated and used to water a nature area, lawns and gardens within the development.

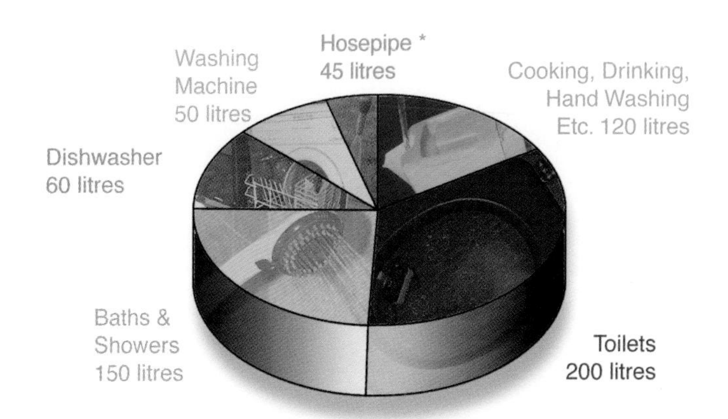

Water use in a typical household of 2 adults and 2 children (total 625 litres = 156 litres per head per day)

THE URBAN WATER CYCLE

In the UK, domestic water consumption is between 140 and 160 litres per person per day. London's 7 million inhabitants use on average 12 882 litres every second (2600 gallons per second) or the equivalent of a medium-sized river.

In most developed countries, water is delivered directly to homes and businesses by public or privately run companies that are licensed to withdraw water from rivers, lakes or groundwater. In England and Wales, approximately two-thirds of the mains water used comes from rivers. The figure is higher in Scotland, Ireland and a number of countries in continental Europe. The remaining third of the water supply comes from water stored in underground aquifers.

Over-abstraction of surface water and groundwater can result in depletion of stream flows and have a detrimental effect on fisheries, wetland areas and the amenity value of

The River Severn at Shrewsbury, with about 13 000 litres/second flowing over Castlegates Weir. This is the same as London's average mains water consumption each second.

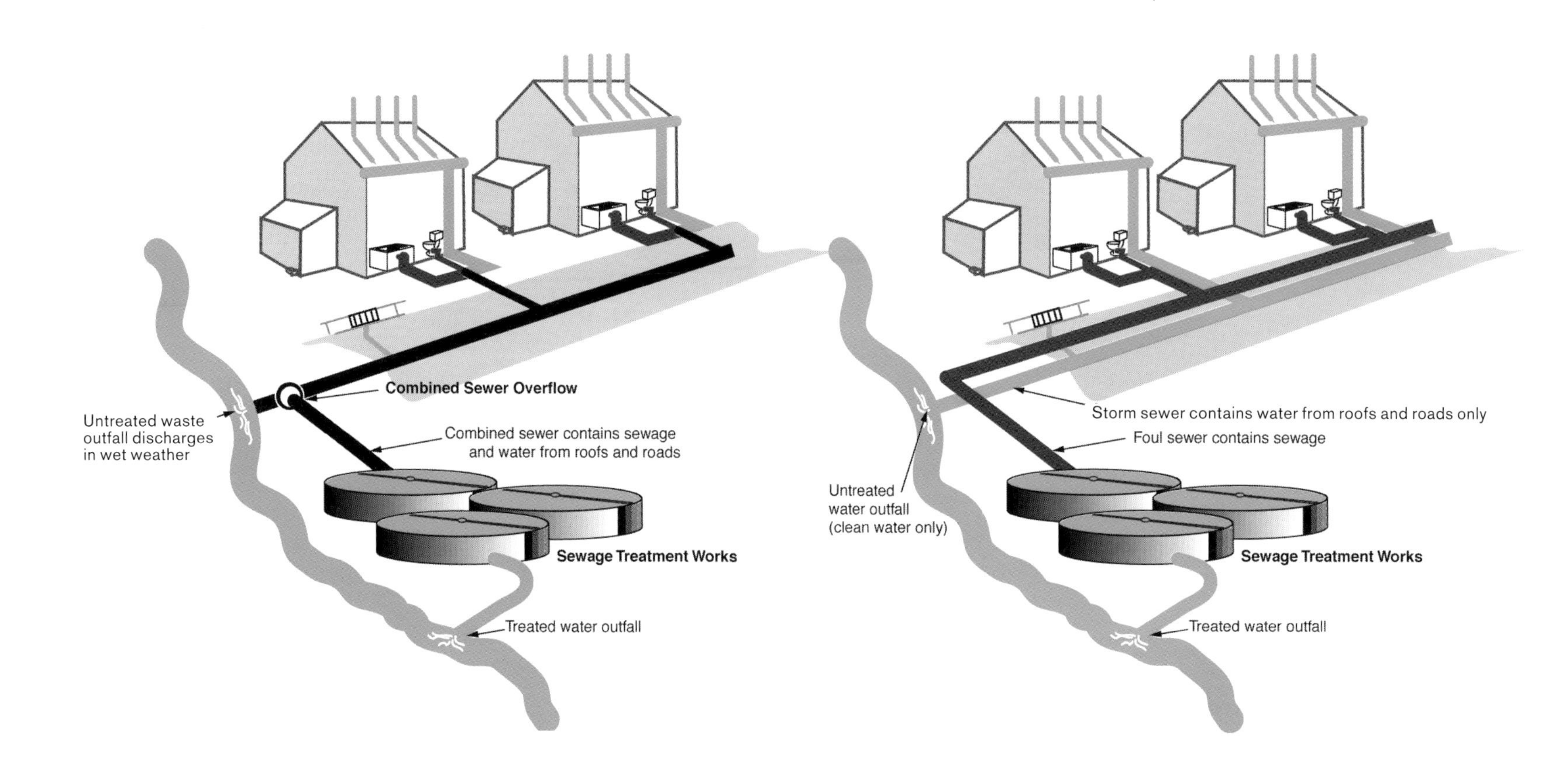

The differences between combined and separate sewer networks.

streams. There are many ways that we can all help to reduce water use – and therefore the volume of water abstractions and waste-water discharges. In the garden, using a hosepipe can consume in an hour as much water as a family of four would use in 2 days! If it is necessary to water, then watering in the cool of the evening means that less water is lost through evaporation. Also, mulches, for example, can be used to help retain soil moisture and savings can be achieved by using water butts to collect rainwater from roofs.
Water used to irrigate gardens is lost to the atmosphere by evaporation and transpiration from plants, but most of the water brought into urban areas via the water-mains network is eventually discharged to sewer, mainly through washing and toilets. The sewer system also collects surface-water runoff from rainwater draining from roofs and other hard surfaces.

There are two types of sewer: combined and separated. Combined sewers take all the water for disposal through a single network of pipes. They collect both sewage and surface drainage from roads and yards, and waste water from industry. This made a lot of sense when there was little or no sewage treatment, and these sewers predominate in older urban regions. However, there are benefits in keeping sewage and industrial effluents separate from surface runoff. For example, the flow in a combined sewer varies a great deal between dry weather, when the flow is all sewage, and wet weather, when it is mostly rainwater. Building large long pipelines to carry mainly rainwater is both impractical and very expensive and the resulting dilute liquid is difficult to treat. Consequently, combined sewers have overflows built into them to limit the flow in the pipe. If the overflows do not operate properly, manhole covers can rise and flood the streets with sewage or pollute rivers by premature operation of the overflow. A major programme of work is now under way to improve the operation of thousands of combined sewer overflows across the UK so that they no longer cause visible pollution or unacceptable water quality.

Unsatisfactory combined sewer overflows can discharge large amounts of offensive litter into streams.

In separated sewers, there are different networks of pipes for rainwater and for sewage and industrial effluents. Separate sewers convey the rainwater through relatively short lengths of pipe direct to the nearest stream, to which

BOX 3.3 **Pollutant types and sources**

The Royal Commission on Environmental Pollution defines the term "pollution" as the result of release to the environment by human action of a substance or energy, that is liable to cause hazards to human health, harm to living resources and ecological systems, damage to structures or amenity, or interference with legitimate uses of the environment.

Pollutants can be divided broadly into a number of principal types, some falling within more than one category.

Suspended solids are of concern in rivers, as deposition may eliminate invertebrate species that prefer clean surfaces. In turbid waters, the reduction in light penetration also restricts growth of water plants. Solids may be organic, as in sewage, or inorganic, as in iron hydroxides (ochre) arising from coal-mine drainage. As well as causing physical effects, suspended solids may also contribute to the biochemical oxygen demand (BOD) load. Many contaminants (heavy metals, nutrients, etc.) are often associated with suspended solids.

Inorganic nutrients – nitrogen and phosphorus – are essential nutrients for the growth of plants including algae. Excess levels of ammonia and nitrate (containing nitrogen) and phosphates (the source of phosphorus) can lead to nuisance levels of plant growth or eutrophication.

Ammonia can arise from sewage, agriculture, fertilisers, aerial deposition and some industrial sources. It is toxic at high concentrations, although the toxicity varies widely with temperature, other water chemistry and, especially, pH.

Degradable organic pollutants arise from sewage systems, food processing, agricultural wastes and some industrial processes. These materials support the rapid growth of a wide range of micro-organisms and tolerant invertebrates, which deplete oxygen from the water. If the pollution load increases beyond the capacity of the river, the concentration of dissolved oxygen decreases and organisms less tolerant of low oxygen levels are eliminated, eventually leaving a community principally comprising bacteria.

Toxic pollutants include a wide range of compounds, from pesticides to industrial chemicals and wastes. Some, such as cyanide and phenol, are readily degradable, provided that dilution is adequate to avoid acute toxic effects. Others, such as many chlorinated organic chemicals (such as dry cleaning fluid) and heavy metals (such as cadmium and lead), are extremely persistent in the river environment, often leading to chronic toxic effects which may only be detected by studying the river biology in detail. The presence of these contaminants in river sediments often leads to problems with river remediation works, where disturbance of the sediments may cause serious pollution.

Micro-organisms, including bacteria, protozoa, parasitic worms and viruses, are all river pollutants associated with sewage and certain industrial effluents. Other sources include the import of animal faeces in surface runoff and large concentrations of birds on urban water bodies. Water treatment techniques are used to disinfect drinking water to make it bacteriologically safe to drink, and waterborne diseases such as cholera and typhoid are now extremely rare in developed countries.

Precipitation of iron hydroxides has turned the bed of the river orange and has killed the river bed organisms on which fish would feed.

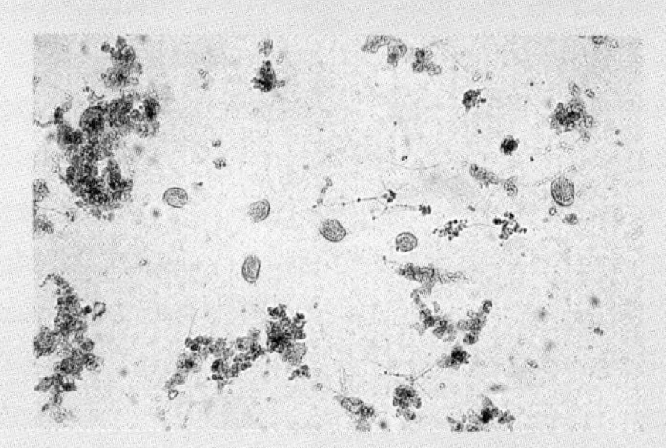

Tiny organisms, like these in this photomicrograph isolated from sewage, are capable of causing a wide variety of serous illnesses.

This storm water drain flowing into the River Nene is contaminated by oil and sewage from a nearby housing estate in Peterborough.

the water would have flowed naturally prior to any development, without treatment. This system has advantages for managing the flow of sewage and trade effluent, which is fairly predictable and therefore a sewage treatment works can be designed to treat it properly.

Over recent years, it has become increasingly clear that the rainwater delivered to rivers via separate sewers is not always clean and can be a major source of pollution of rivers in urban areas. In Scotland, SEPA has shown that 500 km of river have been polluted by the effects of drainage from urban areas delivered to rivers via the surface-water sewer. Shampoo from cars washed in the street, engine oil poured down the surface-water drain (an illegal practice), rubber and metal worn from vehicle tyres and brakes, and sediment and grit washed off roofs and roads, all contribute to pollution in surface-water drainage to streams and rivers.

Urban water pollution

Pollution in urban rivers can take many forms. Some, such as oil and litter, are easy to recognise and generally agreed to be undesirable. Other forms of pollution are invisible but nonetheless dangerous, perhaps rendering the water toxic to natural river life and making it unsuitable for use for water supply and recreation. In Scotland, statistics for 1996 showed that sewage effluent is the primary cause of river pollution, affecting 34 per cent of river length, but urban street drainage is also significant (affecting 11 per cent of river length).

BOX 3.4 **Oil care**

Oil pollution on the River Calder, Yorkshire. (Jenny Hodgson)

Oil (including fuels such as petrol, kerosene and diesel, as well as lubricants) is a frequent pollutant, accounting for a quarter of all notified water pollution incidents in 1999. It is a highly visible pollutant, even when present in small quantities, giving the well-known coloured sheen on a water surface. Oil and fuels damage the environment, wildlife and habitats, contributing to land contamination and stoppages of water abstraction. Much oil pollution results from deliberate or careless actions that can be avoided.

In 1995 the environment agencies of the UK initiated the Oil Care Campaign, together with industrial partners, the objective being to change attitudes and behaviour. The particular targets were the disposal of used engine oil, and the delivery and storage of fuel oils. The involvement of industrial partners led to oil disposal guidance being printed on the packaging of lubricating oils, and in Haynes manuals for the DIY car maintenance market. A helpline provides advice on oil disposal. Many fuel tanks are now manufactured in a way that reduces the chance of a major spill.

Logo for the successful Oil Care campaign.

The campaign has been effective – oil-related incidents have reduced from 6908 in 1994 to 5308 in 1998 – but more remains to be done. The awareness campaign is ongoing, and is soon to be supported by legislation aimed at reducing incidents from fuel storage tanks, by requiring more measures to contain a major spill. The 1000 litres in a small domestic heating fuel tank is enough to pollute 10 km of a river, or to render as much as 80 million m^3 of water unfit to drink (enough for a family of four for several thousand years).

BOX 3.5 **Industrial accidents**

In one of the worst ever pollution incidents, 250 km of a major international river, the Rhine, were devastated on 1 November 1986. Firemen in Basel, Switzerland, called to a blaze at a riverside warehouse owned by Sandoz AG, Switzerland's third-largest chemical manufacturer, flooded the factory with water. The runoff caused 30 tonnes of an extremely toxic cocktail of at least 34 different agricultural chemicals to be released into the river. The results were catastrophic: it is estimated that the incident released as much pollution into the river in an hour as it usually receives in a year, and over 500 000 fish were killed. The spill caused immediate concern for drinking water supplies for the 50 million people along the Rhine (nearly as many people as the entire population of the UK), with many towns in Germany turning off water intakes and bringing in drinking water by truck. In the Netherlands, authorities re-routed the Rhine to speed up its passage to the North Sea, in order to keep the pesticides out of drinking water. Fishing along the river was also banned.

Dead fish are a very visible result of a pollution incident. Fish kills can be caused by domestic sewage, road run-off and farm wastes, as well as by industrial chemicals.

Some pollutants are degraded or taken up naturally in rivers and become a problem only in excessive quantities. Others, such as oils, are not readily broken down and their undesirable properties can persist in the rivers for a long time. Pollutants can enter the river in effluents from point sources: sewage treatment works, sewer overflows and industrial premises. However, diffuse and unidentified sources are equally important and include:

- surface runoff from urban hard surfaces;
- washout of fertilisers and pesticides from managed grassland, parks and gardens;
- sewage and industrial effluent entering surface water drains through wrong connections;
- groundwater contaminated from industrial operations and old landfill sites;
- litter;
- transport accidents;
- illegal tipping of wastes.

During 1999, 30 922 incidents of water pollution were reported to the Environment Agency in England and Wales – an average of over 80 reports a day. Of these, nearly half were substantiated upon investigation. Each incident can cost up to £5 000 to investigate. Of these incidents, 953 were classified as either major or significant. Of these, 37 per cent were caused by sewage and other organic wastes, 23 per cent involved oil and 13 per cent were caused by chemicals.

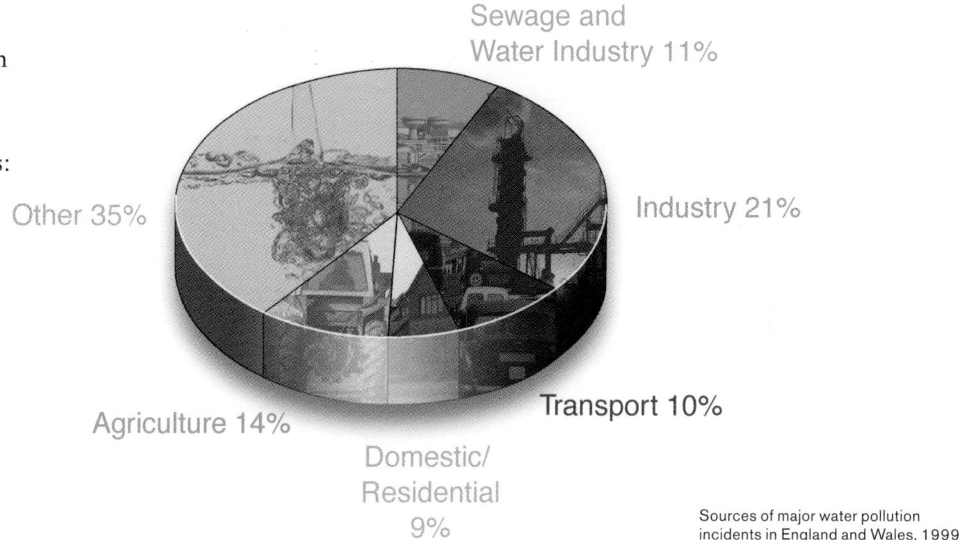

Sources of major water pollution incidents in England and Wales, 1999

The monitoring, assessment and regulation of water quality in rivers is largely based upon chemical measures. However, not all pollutants can be monitored and in any case these measures tell us little of the effects of pollution on living organisms. In many countries including the UK, chemical measures are supplemented by direct biological assessments of the health of animal and plant communities in surface waters. Chemical measurements are instantaneous, describing the river water quality at a single point in time, and require large numbers of measurements for an accurate

BOX 3.6 **Bioassessment of rivers**

Many systems for "bio-assessment" are in use across the world, many of which focus on the small invertebrates that live on the stream bed. These communities of fly larvae, shrimps, snails and worms are present and often abundant in most natural running-water habitats; they are relatively easy and inexpensive to collect, and there are good keys to enable their identification.

In the UK, a Biological Monitoring Working Party (BMWP) was set up in 1976 to provide "an overall view of the condition of rivers and canals and of the discharges to them and to show the effectiveness of pollution control policies". It developed a standardised system for assessing the biological quality of rivers based upon identifying organisms to family groups and then ascribing a score to groups of families according to their tolerance to pollution. Scores of 10 are given to the families most sensitive to pollution, which include many stonefly and mayfly nymphs that are characteristic of clean, fast-flowing, gravel-bed streams. Scores of 1–3 are ascribed to pollution-tolerant organisms such as midges, worms and many snails. The sum of the scores for all families present gives the BMWP score for a site. The average score per taxon is also recorded because it is independent of the number of taxa counted. By comparing the observed values with expected or benchmark figures for each location, it is possible to classify the biological quality of each reach of river and to detect change.

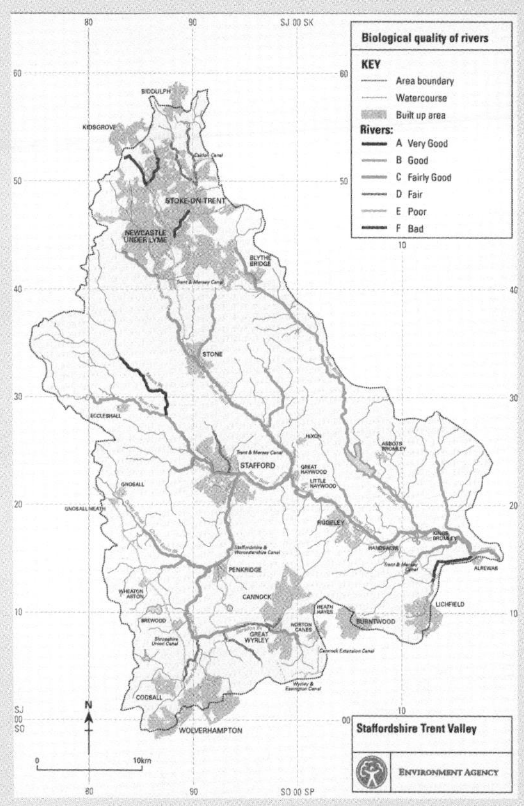

Biological quality of rivers in the headwaters of the River Trent.

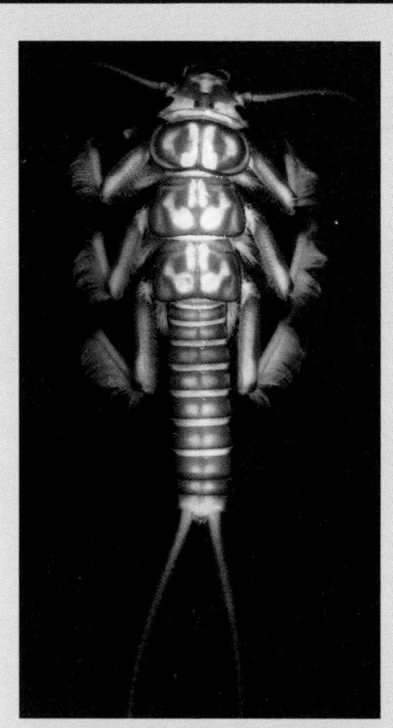

Dragonfly larvae are indicators of a healthy lowland river with a good invertebrate population.

assessment. Organisms respond to environmental conditions over time. They can also show the cumulative effects of many different stresses. Some contaminants can enter watercourses in very low, apparently harmless, concentrations but can be taken up by plants and animals, and accumulated within natural food chains, becoming toxic in higher animals.

The treatment of domestic waste water

Domestic waste water, consisting of used water from bathrooms, toilets and kitchens, can contain a whole range of contaminants. Typically these include biodegradable suspended solids, such as faeces, vegetable and food matter, less biodegradable objects, (toilet paper and some sanitary products), soluble substances (ammonia, chloride and human hormones) and persistent artificial objects, such as panty liners, condoms, cotton buds and plastic wrappers. Other characteristic constituents include soaps and detergents (including phosphates and some metals, such as zinc); household chemicals, such as table salt and bleach; fats, oils and greases; bacteria and viruses; copper from water pipes; and soil and grit.

The objective of sewage treatment is to convert the sewage into water that can be safely discharged into the environment. Each discharge has to meet the standards of a legal consent from the environmental regulator. The consent standards will take into account the ability of the river itself to complete the treatment process, the dilution

BOX 3.7 **Bioaccumulation**

Small organisms at the bottom of the food chain may absorb small amounts of toxic substances. The larger organisms that feed on them may retain the toxic substances, in turn passing them on to large predators that feed on them, at the top of the food chain. Toxic chemicals accumulate in the food chain, a process known as bioaccumulation. Typically, the concentration of a toxic chemical in water may be so low that it appears insignificant, but chemicals can be concentrated at every level in the food chain: plankton, invertebrates, fishes and then birds or mammals. Sometimes the final concentration can be hundreds or thousands of times the original concentration in water.

Significant problems can arise if toxic chemicals continually enter streams and rivers, because over time the cumulative effect of large numbers of small doses can seriously affect the ability of organisms to maintain healthy populations. The story of the magnificent beluga whale in the St Lawrence River, North America, presents a tragic example. Now threatened with extinction, dead whales were found to be so heavily contaminated with a range of persistent chemicals that the carcasses could have been classified as toxic waste! Concentrations of one pesticide were magnified three times in sediment, eight times in invertebrates, 200 times in foraging fish, 350 times in predatory fish and over 2000 times in the whales.

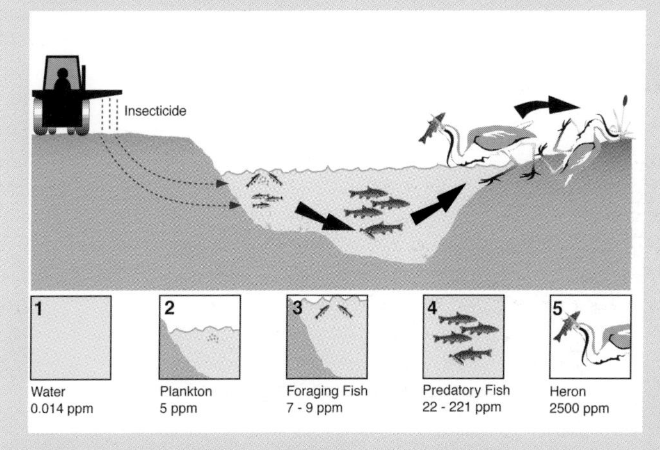

Traces of toxic chemicals may enter water bodies as a side effect of many legitimate activities. Traces of pesticide in the water are incorporated into plankton (1) which form the diet of small fish (2). The small fish contain more pesticide than either the water or the plankton. The small fish are eaten by larger fish (4), leading to further concentration of the pesticide. The larger fish are eaten by birds (5) and the pesticide concentrates even further, more than 100 000x the concentration in the water, leading to the death of the birds.

BOX 3.8 **Sewage treatment**

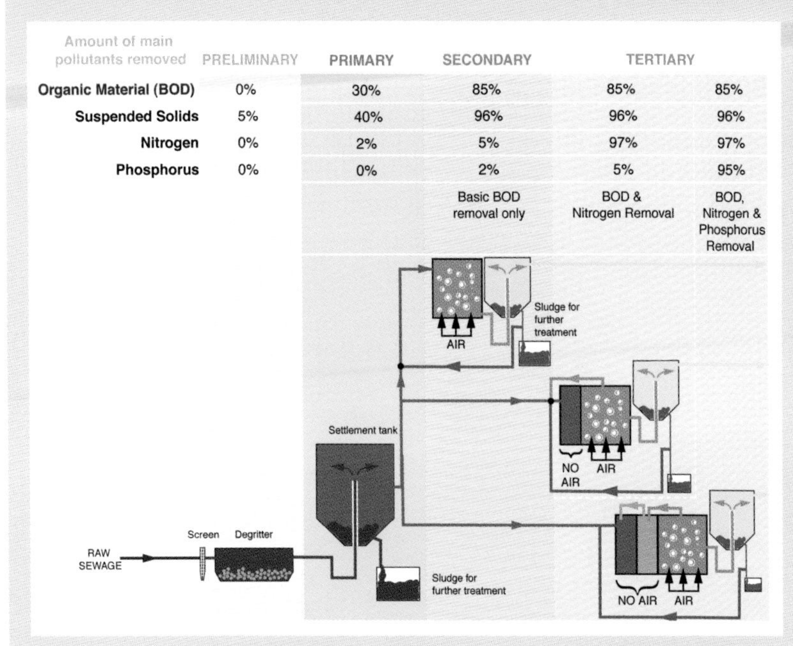

Amount of main pollutants removed	PRELIMINARY	PRIMARY	SECONDARY	TERTIARY	
Organic Material (BOD)	0%	30%	85%	85%	85%
Suspended Solids	5%	40%	96%	96%	96%
Nitrogen	0%	2%	5%	97%	97%
Phosphorus	0%	0%	2%	5%	95%
			Basic BOD removal only	BOD & Nitrogen Removal	BOD, Nitrogen & Phosphorus Removal

The main pollutants in sewage - solids, oxygen demand (BOD), nitrogen and phosphorus - are removed by successive treatment stages.

Preliminary treatment The first stage of treatment is to remove persistent objects and grit by screening and settlement. Screening normally removes all objects greater than 6 mm across. This is carried out to protect sewage treatment processes from blockages in pumps and pipework, thus reducing maintenance costs.

Primary treatment The purpose of primary treatment is to allow as much as possible of the organic material in the sewage to settle out. Large tanks are used to produce the very slow flow that gives time for settlement to occur. Chemicals (known as flocculants) may be added to improve the efficiency of the process, which may remove some 50 per cent of the solids.

In the fixed film process, settled sewage is distributed across porous beds covered with a biological film.

Secondary treatment Secondary treatment relies on a variety of organisms to break down the remaining compounds to carbon dioxide and water, effectively by feeding on the organic matter in the sewage. Biological oxidation may be achieved in one of two main ways: "fixed film" processes such as biological filters and "suspended growth" processes such as the activated sludge process.

The modern biological filter is a circular or rectangular bed, 1.5–3.0 m deep filled with broken stone, slag, or plastic shapes. Bacteria, protozoa, fungi and insects grow on the surface of this solid medium, covering it with a jelly-like film. Settled (primary treated) sewage is evenly spread over the top of the filter and percolates downwards over the surfaces of the medium where the organisms feed on the impurities – and each other. This process relies on the free flow of air through the filter. The effluent draining from the base of the bed is usually clear but will contain some lumps of the active film, known as humus, which must be removed by settlement before the effluent can be discharged to a watercourse. The humus sludge produced usually receives further treatment before disposal.

Activated sludge is generated by the continuous aeration and agitation of a mixture of incoming sewage and sludge from its treatment, so that the sewage and sludge are intimately mixed and have free access to air. The sludge is rich in bacteria and other organisms capable of digesting the impurities in the sewage. After aeration for 6–10 hours the mixture is allowed to settle, producing a biologically stable, clear and well purified effluent. Most of the settled activated sludge is pumped back to the aeration tanks for reuse. The sludge accumulates and some of it must be removed for disposal. About 55 per cent of domestic sewage in the UK is treated in this way. An activated sludge plant requires less land than a biological filter plant (which is important for works processing the waste water from large towns), but is more expensive to operate.

Tertiary treatment: The residual organic content and suspended solids from a secondary treatment process can be further significantly reduced by tertiary treatment, which can also remove phosphate and nitrate. Tertiary treatment systems may use constructed "natural" systems such as lagoons or reedbeds, or equipment such as rapid sand filters.

Sewage sludge

Sludges from sewage treatment are usually first thickened, a process of partial drying to reduce their volume. Digestion further reduces the volume and produces methane gas, which is burned to produce heat for the treatment plant and perhaps electricity for export to the national grid. The digested sludge may then be used as fertiliser on agricultural land, or in industrial processes.

In the activated sludge process, sewage, air and micro-organisms are mixed vigorously. (Entec)

available in the river, and the uses and amenity value of the river. Sophisticated sewage treatment is expensive, and so there needs to be a proper balance between cost and benefit.

Since the contaminants in domestic waste water are mainly organic in origin, a typical waste water treatment plant is geared towards the biological breakdown of the organic matter into carbon dioxide and water. It is also necessary to remove solids that cannot be degraded by biological oxidation. In addition to treated effluent, the sewage treatment process produces solid waste – the non-degradable solids in the sewage – and a sludge resulting from the biological oxidation.

There are four levels of treatment. In the UK, 76 per cent of the population currently benefits from the highest two levels: secondary and tertiary. Only 11 per cent are limited to a minimal, or preliminary, level of treatment. Planned improvements will enable the majority of urban works to deliver secondary or better levels of treatment by 2005. According to the level of treatment there will be a substantial reduction in the major pollutant characteristics. Typically 70–90 per cent of the organic matter that is responsible for the oxygen demand and suspended solids of sewage will be removed. Ammonia, heavy metals and most other potentially toxic domestic chemicals will also largely be removed and there will be some reduction in the nutrients nitrogen and phosphorus. Certain substances such as salt remain unaffected. There will also remain a bacteriological load associated with the waste water.

BOX 3.9 **Chicago's recovering waterways**

In 1885, Chicago was devastated by an epidemic of typhoid, cholera and other waterborne diseases, which killed 100 000 people. Sewage treatment facilities were introduced from 1919, and by 1922 the flow in over 300 km of rivers, creeks and canals was reversed and made to flow away from Lake Michigan, to the Des Plaines River. Between 1984 and 1993, 150 km of tunnels were constructed to intercept contaminated storm waters.

To manage the quality of the water in the streams and rivers of Chicago, artificial aeration was introduced in 1972. Before the dissolved oxygen concentration becomes so low that it is harmful to aquatic life, in-stream aeration transfers oxygen to a waterway, by mechanical or other means. The first aeration system passed compressed filtered air through porous ceramic diffuser plates placed on the bed of a channel. Subsequently a second aeration design, the Sidestream Elevated Pool Aeration, was introduced. This involves high-volume, low-head pumping of a portion of the streamflow with relatively low dissolved oxygen levels from a channel by means of screw pumps, to a series of elevated shallow sidestream pools linked by waterfalls. As a result of this integrated approach to pollution management, between 1975 and 1995 mean dissolved oxygen concentrations increased by about 59 per cent. Mean ammonia levels decreased by over 75 per cent. The number of fish species increased from 7 to 16, including a number of game fish, and the total number of fish twenty-fold. The Chicago waterfront and its riverside parks have become important attractions in the regeneration of the city.

In-stream aeration is used to improve water quality and make attractive park features in Chicago.

BOX 3.10 **Using waste water**

Essex is the driest county in the UK, and half its water is imported from other areas. Continuing growth in household numbers and uncertainty about impacts of climate change are raising increasing concerns over the reliability of existing water resources. An attractive new supply option is the reuse of Chelmsford's treated waste water, so boosting local resources for 1.4 million people by 10 per cent. The plans involve intercepting a pipeline carrying the waste water to the coast and providing substantial additional treatment, before discharging the treated water into the River Chelmer to support an abstraction about 4 km downstream. The consent to discharge to the Chelmer requires the discharge to cause no deterioration to the existing water quality, and necessitates the reduction of phosphorus and nitrogen levels in the effluent. A final UV treatment stage will be provided to secure a microbiological quality to help protect recreational users of the river. Initial concerns about oestrogenic properties of the mainly domestic waste water were overcome when research demonstrated that the compounds involved are broken down by such treatment as ozone and activated carbon at the drinking water treatment works.

Industrial waste waters

Industrial waste waters can be more complex and highly variable in nature than domestic waste water. In addition to the components of domestic sewage, industrial effluents may contain specific toxic and hazardous substances, such as organic chemicals including solvents and dyes, metal compounds, and oil. Some industrial effluents contain similar substances to domestic waste water, but in larger quantities, such as waste water from many food and drink factories.

In many urban areas, industrial waste water goes into the same sewers as domestic waste water but must comply with trade effluent consents issued by the sewage undertaker.

Trade effluent consents control industrial effluent discharged to sewers. This is in order to prevent the effluent causing damage or harm to the sewerage system, and to ensure that nothing is discharged that cannot be treated by the waste water treatment works to a satisfactory standard. The water company involved will set maximum allowable limits for certain toxic substances in the effluent (ammonia, heavy metals, etc.), which may affect the normal treatment process, and for acids and volatile chemicals, which may attack the sewer system or endanger the health and safety of personnel. In addition, a trade effluent consent requires the industry to pay an appropriate charge to the sewerage undertaker for the conveyance, treatment and disposal of the effluent. The industry must also provide data on volumes and

Large fossil fuel power stations have cooling towers to prevent thermal pollution but consent standards must be met before discharge to watercourse or sewer. This is because the evaporation of water that produces the cooling leaves behind the natural salts in the water, and chemicals may also be added to the circulating water to prevent the growth of algae that would block pumps and Legionella bacteria that could infect humans (River Severn at Ironbridge).

BOX 3.11 **Cleaning up the River Tees**

Northumbrian Water's Bran Sands TEES effluent scheme is intended to make a major contribution to cleaning up the River Tees, formerly one of Britain's most polluted large rivers. The effluents to be treated derive from the large Teeside conurbation and waste waters from the chemical, and iron and steel industries. The treatment works will be the largest industrial and municipal effluent plant in the UK. The plant, completed in December 2000, treats 300 000 m^3/day of effluent, a load equivalent to a population of 3.5 million.

The £140 million Bran Sands scheme, built on a brownfield site, combines sewage treatment with the treatment of a number of industrial effluents from companies such as ICI, Du Pont and Corus (formerly British Steel). In addition a £70 million regional sludge drying centre on the same site handles the sludge from Bran Sands and other major sewage treatment plants in the Northumbrian Water area. It is proposed that the centre will eventually produce fuel pellets for a combined heat and power plant that will enable the entire complex to be self-sufficient in heat and electricity. The TEES scheme offers a potential long-term strategy for sustainable treatment of sewage and industrial effluent for the North East of England and importantly will help to in re-establish the Tees as a major migratory fishery.

Now that the water in the River Tees is cleaner, it can again be used for recreation involving contact with the water. The Tees Barrage, Middlesborough. (John Hunt)

character of discharges, for use in the design of future sewage systems and sewage treatment works.

A significant waste discharged to rivers by some industrial processes is heat. When water is used for cooling, the water discharged to the river is warmer than that abstracted. Warm water holds less oxygen in solution than cold water and therefore warm water can cause distress to fish, tending to suffocate them. For this reason limits are usually placed on the amount of warm water that can be discharged, depending on the river temperature downstream of the discharge.

Industrial waste water treatment

Many industries produce effluents that are not directly suitable for treatment by a sewage works designed for domestic waste water. In these cases the following options are possible.

Firstly, the effluent can be specially treated on site to achieve a quality suitable for discharge to sewer and subsequent treatment at a conventional sewage works. Although industrial waste waters have a wide range of characteristics, the wastes from particular processes are often well defined. It is therefore possible to design specific treatments for these wastes, producing a treated waste water that is acceptable for discharge to sewer for co-treatment with domestic sewage. Examples of pretreatment include screening or filtration to remove solid materials; neutralisation of acids or alkalis; removal of heavy metals,

Northumbrian Water's TEES effluent treatment plant, under construction.
The plant has been built to treat sewage from Middlesborough and effluent from the heavy chemical industry of Teesside, which can be seen on the skyline across the River Tees. (Entec)

oils and solvents or partial removal of biodegradable organic matter. In some cases, this treatment will be so effective that the water is suitable for direct discharge to surface water, or for re-use as process water in the factory.

Where industrial effluents are difficult to treat alone, they may be discharged to a specialist works designed to co-treat domestic and industrial effluents. Here, the domestic effluent supplies essential nutrients for the organisms that will decompose the industrial waste. The industrial effluent is piped to the treatment site by dedicated pipelines where it is received and pretreated by methods as above, prior to being mixed with domestic sewage for co-treatment.

BOX 3.12 **The future of waste water treatment**

The development of biological processes to remove nitrogen and phosphorus from treated waste water introduced biochemical pathways that were previously unknown to sewage treatment scientists and engineers. It also opened the doors to the development of significantly more economical and stable activated sludge systems, minimised secondary environmental impacts and reduced the cost of implementation. Nitrate contained in the sewage is used to oxidise some of the other pollutants; the energy costs of mixing air into the sewage are reduced and the volume of sludge produced is also smaller.

However, the process is slow and the reactor vessels have to be large. Size matters for a large works and the trend is towards high-rate technologies that enable the works to take as little land area as possible, and to operate with very low manning levels. In some cases the works need to be totally enclosed because of proximity to other developments, and sophisticated ventilation systems are required to prevent odour nuisance and to ensure a safe working environment for the operators.

Treatment plants are being made more compact by increasing the concentration of suspended bacteria – the more there are, the faster the job is done. The latest idea is to use membrane filters to separate the sludge out, permitting operating bacteria concentrations of 15 000 mg/l – three times maximum concentrations using standard techniques – producing extremely high-quality effluents and reduced sludge volumes for treatment. Significant costs are involved in installing and cleaning the filters and this technology, developed in Japan, is currently more expensive to operate than conventional plant. However, plants are running in Wales and Ireland and improved designs are likely to become economically attractive.

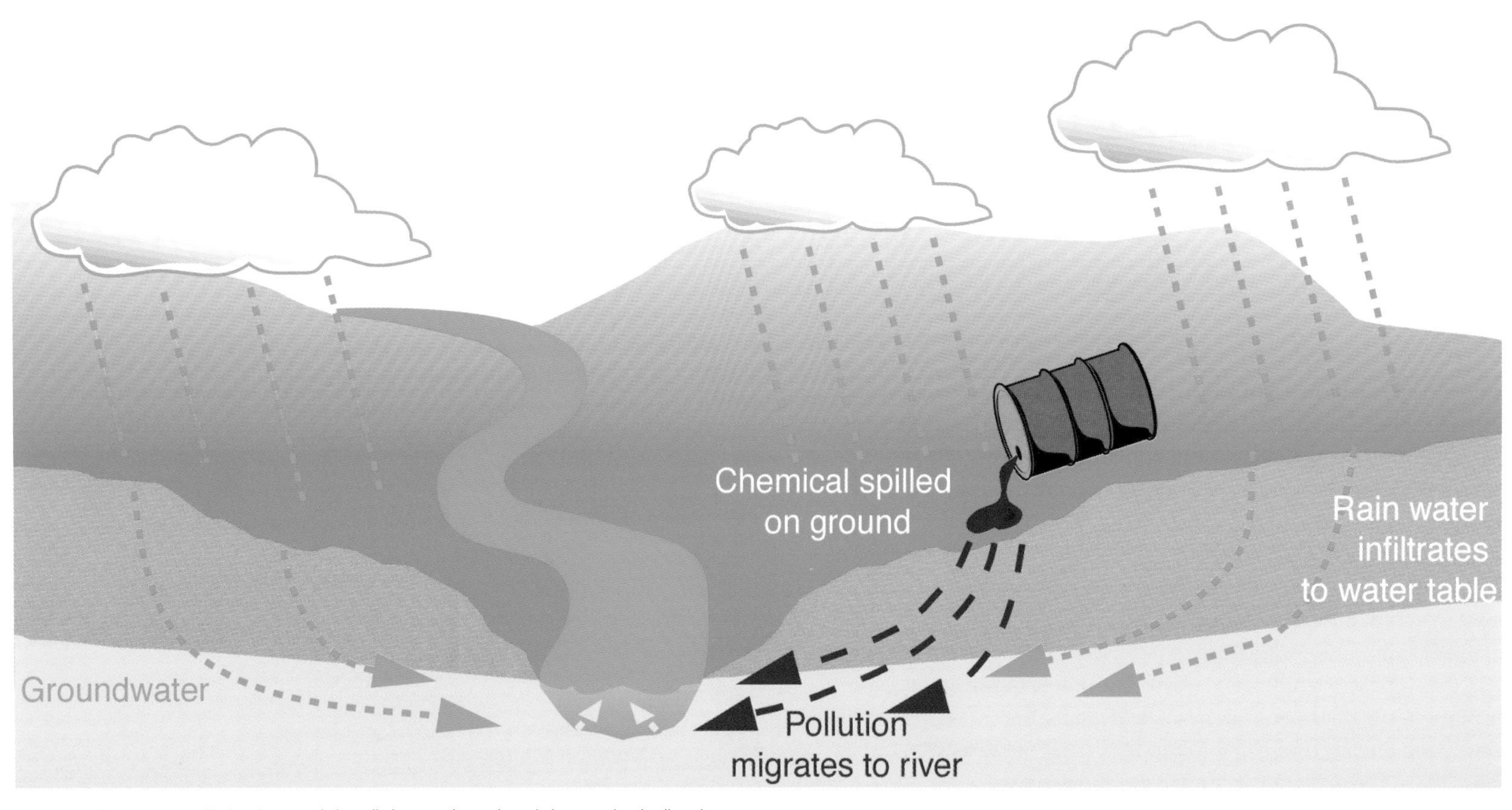

If polluting substances are spilled on the ground, the pollution can migrate through the ground and pollute rivers.

Diffuse pollution

Some pollutants are washed off the soil and urban surfaces, or seep through the ground, often from a very large number of small sources. The cumulative effect of all of these sources can cause a general degradation in urban water quality even though there are no obvious concentrations of pollution (point source discharges). The significance of these diffuse sources of pollutants has become more apparent because of the considerable success in reducing point source pollution. Groundwater seepage is the natural source of summer river flow. Polluted groundwater may arise both from industrial contaminated land and from "green" land such as parks, gardens and playing fields.

Green spaces in urban areas

Most urban areas have at least some green land, comprising parks, private gardens, allotments, playing fields, golf courses, traffic islands, and road verges and central reservations. The application of chemicals as fertilisers, herbicides (to control weeds) and insecticides (to control pests such as greenfly) causes problems in these areas. The way the chemicals are used results in large proportions of them reaching the ground surface. Here one of three things may happen: rain may wash the chemicals into the nearest drain or watercourse; the chemicals may soak into the ground where they may degrade or enter groundwater; or in the case of fertiliser they may be absorbed, at least temporarily, into plants.

Surface runoff can carry contaminants into rivers, particularly nitrates and phosphates from fertilisers, herbicides and insecticides. Chemicals that soak into the ground can reach groundwater, which may eventually contaminate rivers. To combat this problem, a number of chemicals have been designed either to migrate very slowly (for example paraquat) or degrade very quickly (many lawn weedkillers). In this way the substance has become harmless before it reaches groundwater. Unfortunately, this does not always occur and some chemicals, such as nitrate in fertiliser, do not degrade. Proper use of these chemicals, in accordance with the manufacturer's instructions, can greatly reduce their impact.

Urban green spaces can lead to rivers being polluted by fertilisers and pesticides. (Geoff Dowling)

Contaminated water seeping from the ground, resulting in surface water pollution. (Jenny Hodgson)

Contaminated land

Land may be contaminated by naturally occurring toxic minerals in the soil. More usually, contamination occurs as a side-effect of use, for example, by spillage of chemicals in a manufacturing process, or by the deliberate disposal of wastes. Accidental leakages from storage tanks and pipes, and spillage from fires and transport accidents can also lead to land contamination. Almost all chemicals are sufficiently soluble in water to make the water unfit to drink, even substances like petrol and dry-cleaning fluid that appear not to mix with water. Even small quantities of some substances can cause severe problems. For example, one teaspoonful of dry-cleaning fluid is enough to make a swimming pool full of water unfit to drink. Although the disposal of wastes is now strictly controlled across Europe, the present legacy of contaminated land suggests that careless disposal happened frequently in the past.

Most waste produced in Britain, both domestic rubbish and commercial waste, is eventually disposed of to landfill – in holes in the ground, often former quarries. Disposal to landfill can lead to contaminated groundwater discharging into rivers, in the same way as industrially contaminated land. This is a particular problem with old and poorly constructed landfill sites, although more modern landfills are now designed to prevent contamination. As the waste decays, an offensive leachate comprising a strong-smelling liquid rich in ammonia, and with a large biochemical oxygen demand, can seep out of the bottom of the waste. If industrial wastes are disposed of to landfill, toxic chemicals may also be present. Although leachate contaminated with toxic chemicals is rare, co-disposal of hazardous liquid waste with domestic rubbish is being discontinued as a result of changes in European law.

Urban rivers can be protected from pollution arising from contaminated land in a number of different ways. In the first instance, the land itself may be cleaned up by treating or removing the contaminated materials present. This has the advantage of allowing the land itself to be brought back into

BOX 3.13 **Activities with the potential to contaminate the ground**

Activity	Substance	Properties
Gas works (once present in most towns)	Tar	Causes cancer in humans
	Ammonia	Toxic to fish
	Cyanide	Toxic to most life
	Phenol	Corrosive to skin, toxic to fish, foul taste in drinking water
	Sulphur	Inflammable, damages concrete
Dry cleaning, clothing and computer manufacture	Chlorinated solvents	Causes liver damage, cancer in humans
Motor vehicle fuelling and repair	Petrol	Explosion risk, bad taste, toxic
	Diesel	Bad taste
	Lubricants	Bad taste
	Antifreeze, hydraulic fluid	Toxic to humans
Metal goods manufacture (for example nuts and bolts, cars, electrical appliances)	Chlorinated solvents	Causes liver damage, cancer in humans
	Heavy metals	Toxic, carcinogenic to humans, prevent plant growth
	Cyanide	Toxic to most life
Chemicals and pharmaceuticals manufacture	Chemicals	Most are toxic and many cause cancer in humans
Timber preserving	Creosote	Causes cancer in humans
	Copper	Toxic to animals, kills plants
	Chromium	Causes skin disease in humans
	Arsenic	Causes cancer in humans, cumulative poison
Leather manufacture	Chlorinated solvents	Causes liver damage, cancer in humans
	Chromium	Causes skin disease in humans
	Pesticides	Toxic, especially to fish

This stream was being affected by pollution seeping from beneath a former gas works. This site is now being cleared to remove the source of pollution. (Jenny Hodgson)

Many modern landfills are lined with thick plastic sheeting to prevent leachate from the waste escaping and polluting groundwater. (Entec)

Oily contamination from a former chemicals factory bubbled up through the bed of the River Calder in Yorkshire until the Environment Agency undertook remedial works. Note the characteristic coloured sheen on the water. (Jenny Hodgson)

beneficial use. The traditional approach has been to remove the contaminated soil and other solid materials (such as brick, concrete and fly ash) to a licensed landfill, often some distance away in the countryside. This is increasingly seen as merely transferring the problem from one place to another and over the past decade we have sought more sustainable solutions.

Many different new technologies are now emerging for treating contaminated soils. These technologies seek to destroy the contaminants, to separate and remove them from the ground, or to reduce their pollution potential. Most of the new techniques can be applied on a contaminated site, and some can work with very little ground disturbance, minimising dust, noise and odours associated with excavating contaminated soils.

Where the contaminated ground itself cannot be restored, it is possible to prevent the pollution reaching the river by creating a barrier to water flow. This may be achieved by placing a waterproof barrier on top of the ground like a cap, which reduces the amount of rain entering the ground, thus reducing the amount of contamination dissolved out. A cap can be engineered from materials such as clay and specialist plastics, and certain land uses such as car parks and roads can also form effective caps. Other waterproof barriers can be placed within the ground like a wall, often made by using a slurry of clay and cement which subsequently hardens, or by piling steel sheets into the ground. Since the contamination remains, use of the land may be restricted but many high-value urban developments, such as shopping centres, can be built on land where contamination is controlled in this way. The potential for contaminated land to pollute urban waterways is considerable, but it must also be set in context with other sources of pollution such as industrial discharges and road runoff. Although the problems of contaminated land can be effectively tackled, some sites will only be cleaned up over a long period of time and at a high cost. Prevention is better (and cheaper) than cure when dealing with ground contamination, and while we manage the problems created by our industrial legacy, we should ensure that we create no new areas of contaminated land.

A wall of cement-bentonite slurry is being emplaced in the ground to prevent ground contamination moving out from a contaminated site. (Entec)

Two Separate Systems
Two Separate Tasks

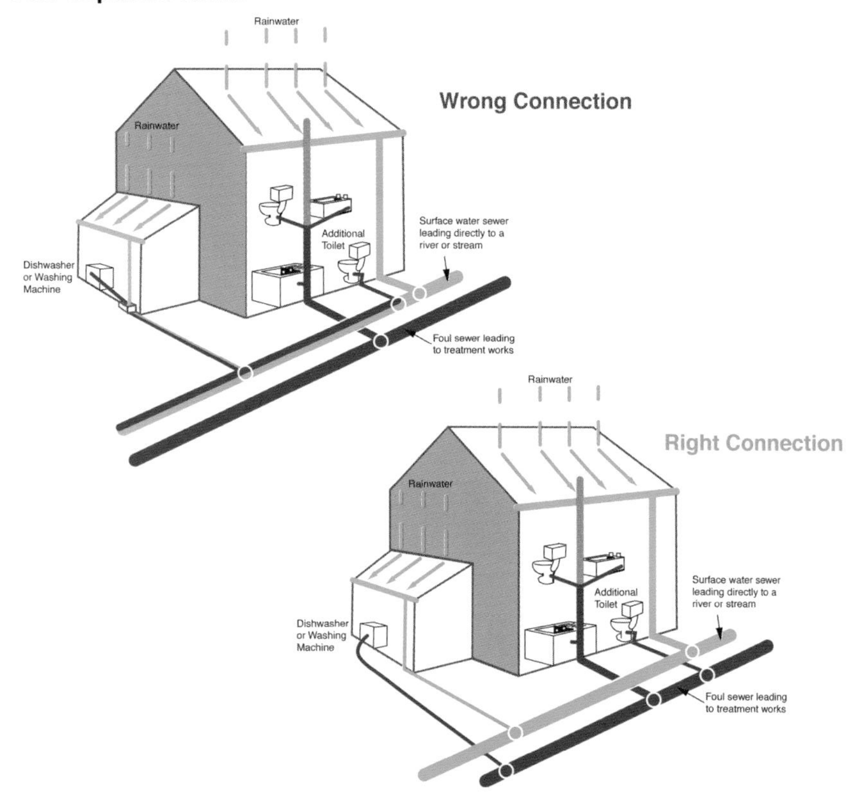

When houses are served by separate sewer systems, it is essential to connect drains to the correct sewer, otherwise streams may be polluted with sewage (Northumbrian Water plc).

Wrong connections

Storm drains are intended to carry clean rainwater quickly to the nearest surface watercourse. Therefore, there are no mechanisms built in to treat this water. Surveys have shown that up to 20 per cent of all properties have incorrectly connected drains, over 70 per cent of these being privately owned houses and 17 per cent publicly owned housing. Most of the problems with domestic properties arise when new facilities are added, a washing machine, dishwasher or new sink that is connected to the nearest drain, which is often the surface water drain from the roof. These connections are illegal and may cause gross pollution as water from washing machines and toilets is discharged untreated into small streams.

A survey of England and Wales in 1999 showed that there are around 1 million wrong connections of this nature, but the number may be as high as 3 million. Additionally, there are around 300 000 wrong connections associated with business premises, where there is potential for more harmful wastes to be discharged untreated into surface waters. Between 2–5 per cent of all pollution incidents investigated in England and Wales are caused by wrong connections. The responsibility for avoiding this kind of pollution rests with the owners of individual properties, but water utilities will help owners identify the correct drain to which new facilities should be connected.

BOX 3.14 **New water quality problems**

There are serious concerns that human impacts upon surface waters may increase risks to human health. A wide range of pathogenic organisms (microbes that cause illness in humans or animals) are found in river water but rarely in sufficient numbers to cause infection. For example, pathogens that infect the gut may be released directly by animals defecating close to or in rivers, by accidental runoff of slurry from farms, by treated sewage effluent discharges from sewage works, by untreated sewage discharges from combined sewer overflows, or by contaminated runoff from roads, gullies, surface water drains, etc. The degree of pollution by animal and human faeces can be measured by tests for certain faecal bacteria for which there are EC Bathing Water Directive standards. Such tests show that most urban streams and rivers will be more contaminated than the worst contaminated bathing waters in the UK.

Historically, outbreaks of typhoid and cholera were caused by the contamination of drinking water with human faeces. Modern water purification treatments and sewage systems have controlled disease outbreaks in the developed world, although problems can still occur in less developed countries. Although conventional water treatment disinfection processes will deal effectively with bacterial and viral pathogens, protozoan parasites such as *Cryptosporidium* will survive through normal treatment processes. *Cryptosporidium* has been the cause of several major outbreaks of disease in the UK and is now routinely monitored in drinking water supplies.

In animals, many of the body's functions, including growth, development and reproduction, are regulated by a complex network of glands known as the endocrine system. The glands release chemical messengers or hormones directly into the bloodstream, where they are transported to tissues and organs throughout the body. The hormone molecules bind to receptors within the tissue and trigger a response, for example, to produce a protein or stimulate cell division. The endocrine system is complex and may be disrupted by both natural and synthetic chemicals found in the environment. These include natural hormones, the synthetic contraceptive pill and nonylphenol, a degradation product of some surfactants (found in detergents). A range of the active ingredients found in herbicides and pesticides are also suspected of acting as endocrine disruptors, as are several phthalates (used as plasticisers in flexible plastics).

Evidence for endocrine disruption has been found in wild populations of fish, birds and reptiles, which have shown changes in the development and function of the reproductive system and the growth and organisation of tissue. Concern about the apparent "feminisation" of male fish was raised in the late 1980s when intersex (showing both male and female characteristics) fish were discovered in water lagoons exposed to sewage treatment water. A national survey in 1998 revealed widespread intersex conditions in male fish in eight river systems. The major contributor appears to be natural and synthetic oestrogens within treated sewage effluent discharged to those river systems.

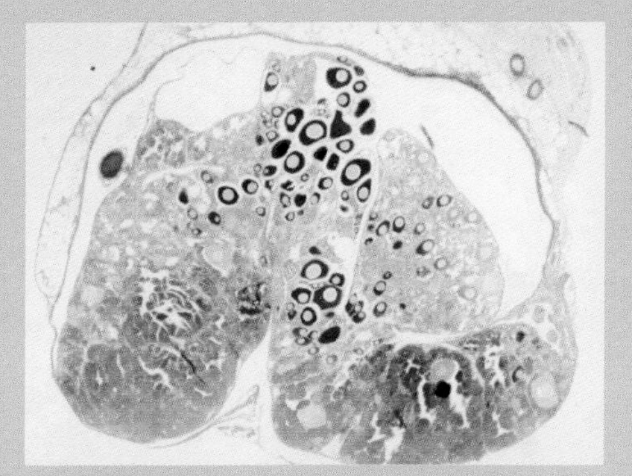

Low-power photomicrograph of section through testis of a partly-feminised male roach from a UK lowland river. Note many small primary oocytes (developing eggs) (small dark-stained circular structures) within normal testis tissue. A large ovarian cavity is visible below testis lobes, and sperm duct is completely absent.

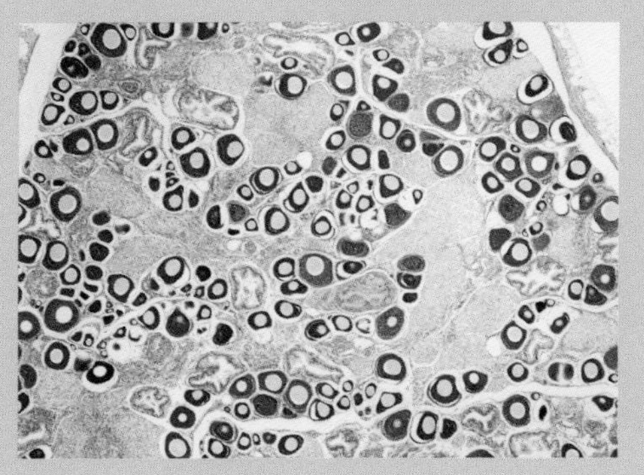

Moderate-power photomicrograph of a section through testis of a partly-feminised male roach from a UK lowland river. Note many primary oocytes throughout testis. Most tissue is female in nature, a more severe example of the intersex condition.

BOX 3.15 **Sustainable urban drainage systems (SUDS)**

Porous brick paviours allow rain falling on a car park to soak in, rather than becoming troublesome surface runoff.

A swale provides a satisfactory means of managing runoff from roads.

The system of separate storm drainage that was installed in many developments in the latter part of the 20th century has been found to have two major disadvantages. The rapid transmission of rainfall runoff into surface watercourses contributes to urban flooding and damages the natural ecosystem. In addition, surface-water drains take pollutants directly to a watercourse. Continued installation of such systems will only make the problem worse. A new approach is now being taken which, by mimicking the natural processes in a river catchment, aims to be more sustainable. This involves the imaginative use of a range of techniques, known collectively as Sustainable Urban Drainage Systems or SUDS.

Key elements in all SUDS schemes are to:

- avoid runoff if possible (allowing it to infiltrate the ground);
- slow the runoff process (which includes the provision of additional storage);
- manage runoff as close as possible to the point of origin (that is, keep things small and local).

Wherever possible surface runoff should be avoided by using permeable surfaces. In addition to gravel and grass, it is now possible to make hard surfaces, from brick paviours to tarmac, that allow water to flow through them. In many cases, where the pollution risk to the aquifer is acceptable, the water can then recharge the groundwater. Where possible, water from roofs can be discharged to soakaways for each house, again preventing it from rapidly becoming surface water. Even road drains can be discharged to soakaway, although the risk of groundwater pollution must be duly considered, since road runoff can contain toxic substances from tyres, spilled oil and residues from fuel combustion. Where the water will not infiltrate, perhaps because of a heavy clay soil, the water from beneath a permeable surface will have to be discharged to surface water. But soaking through the permeable surface introduces a time delay which is valuable in reducing flooding risk. Surface water drainage systems can be designed to include small-scale storage, ranging from broad ditches (swales) and flat areas that are usually dry, to wetlands and small ponds. Wetlands and ponds can be designed to be attractive elements of the urban landscape, thus producing a double benefit.

In addition to slowing the water down, reducing the risk of flooding and damage to river life, these structures provide time for sediments to settle out and for biological processes to break down some of the pollutants. Permanent ponds and wetlands with shallow water and numerous water plants are the most successful in removing pollutants, but there may not always be space for these.

During 2000, CIRIA compiled and published design guidance and good practice reviews to help promote SUDS techniques.

Urban flooding is costly and inconvenient (River Quaggy, North London). A flood relief scheme has now been constructed to avoid a repeat of this incident, using recreation spaces to store excess water.

The screen on this drain is almost completely blocked with rubbish, leading to water backing up in the drain and flooding the street.

NEW STRESSES ON URBAN RIVERS

In addition to increased pollution, rapid urbanisation and industrial growth have introduced other new stresses to rivers in urban catchments, leading to an increased risk of flooding, deterioration of water quality and loss of recreation and amenity uses.

Increased flood risks

Flooding in inland urban areas has two causes: failure of the urban drainage network to remove rainfall fast enough, so that it accumulates; and flooding by an adjacent river as a result of rainfall in the catchment upstream. The two processes may occur together, and cause unusually severe flooding. Even curing flooding in one part of a catchment through improved drainage can make problems worse downstream. In coastal areas, flooding may also occur as the tide backs up.

River flooding can be increased by land-use change throughout the catchment. Rain runs off bare winter fields far more quickly than it runs off well-vegetated areas, such as native woodland. In the long term, flood management should include land-use management, especially woodland, wetland and flood plain restoration, within comprehensive catchment management plans.

Local flooding

Rainfall in urban areas becomes runoff within minutes of the rain starting, as there are few permeable surfaces into which it can soak, even temporarily. The urban landscape, with its roofs, asphalt roads, and paved and concrete surfaces, is designed to shed rainfall into the nearest drain as quickly as possible. This water is then piped directly to the nearest stream, causing a very rapid increase in flow and corresponding rise in water level. The flow rate down surface-water drains may vary considerably, possibly falling to nothing during dry weather but increasing to a torrent during rain. Surface-water drains typically have no spare capacity when the flow exceeds the design amount. Whereas a natural river may expand on to the flood plain, a surface-water drain can only overflow on to the streets. The problem of limited capacity within existing surface-water drains may be further exacerbated by debris and litter that collect in the drains, impeding the flow.

In addition to flooding caused by unusual rainfall events, the nature of many surface-water drainage systems makes them vulnerable to failure. Because of the high density of roads in urban areas, the storm drainage system involves frequent culverts. These culverts can become blocked, especially where they are fitted with entrance and exit grids to discourage trespass. Once obstructions, ranging from crisp packets through to shopping trolleys, Christmas trees and mattresses, begin to gather at the entrance to a culvert, flooding upstream is inevitable.

The severity of urban flooding can be alleviated by reducing the volume of surface runoff. Wherever possible, surface runoff can be avoided by allowing the water to soak into the ground, either through permeable surfaces or through soakaways. When runoff cannot be avoided, the water storage areas are necessary to reduce the peak flows that cause flooding. These ideas have been considered in the development of sustainable urban drainage systems (SUDS).

SUDS must be included at the design stage to try and maintain as far as possible the natural runoff regime and to prevent flooding downstream of the new development. However, finding areas to store water in already urbanised catchments can be difficult, and a mixture of on-site storage and strategic downstream storage may be necessary. Storage can take several forms, including underground tanks, specially designed swales, basins or ponds. Storage can be generated on artificial flood plains, areas of land low enough so that occasional flooding can occur without severe consequences. Recreation areas may meet this requirement – a football pitch that floods once a year is less inconvenient than a flooded house!

Large river flooding

Many of our major towns and cities are situated on the banks of large rivers, for important historical reasons. However, it is the nature of major rivers to flood. In the early days of towns, buildings tended to be confined to higher ground and the flat land adjacent to the rivers was used for grazing. As towns expanded, this flat flood plain was used for development and

Swale, Ure and Ouse area Main river and washlands

Areas of washland extending many miles upstream help to prevent York from flooding.

The November 2000 floods in York, showing the start of the flooded washlands upstream of the city. (Peter Smith Aerial Photography, Malton)

BOX 3.16 **Maidenhead Flood Relief Scheme**

When Maidenhead was severely flooded in 1947, by what was estimated as a 1 in 56-year flood, 120 people were evacuated from their homes. A flood of similar size nowadays would affect over 12 000 people directly and cause damage estimated at £40 million. The natural channel of the River Thames at Maidenhead has a capacity of 285 m^3/s; flows higher than this result in the river overtopping its banks. As part of a flood alleviation scheme, an artificial channel was designed to carry additional water. The channel is 45 m wide from bank to bank, producing a combined capacity of 515 m^3/s, equivalent to a 1 in 65-year flood and more than sufficient to manage the 1947 flood. Care has been taken to ensure permanent flow in the new channel, to obviate many of the aesthetic and ecological problems that may otherwise be associated with flood relief channels.

The new relief channel under construction in 1999. The channel is excavated down to the water table, so that it forms a permanent water feature. (Realistic Photography)

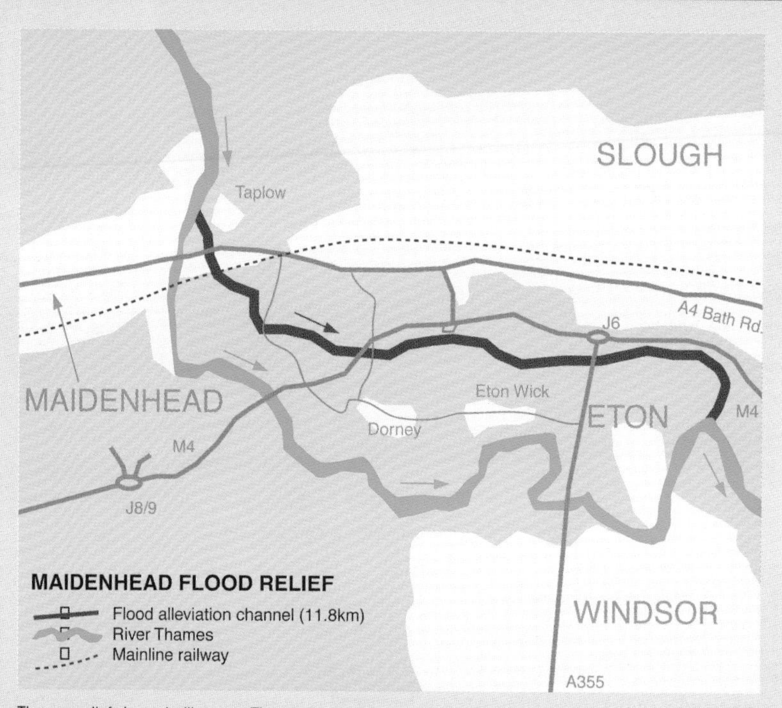

The new relief channel will convey Thames floodwaters round the Maidenhead area.

flooding of this land became a considerable nuisance, with a high cost in terms both of money and disruption.

Water storage has a significant role in reducing flooding, by delaying the influx of water so that it can flow within the channel capacity. Storage can be generated in permanent reservoirs but is more usually over washlands, where the flood plain's natural function – to store flood waters – is restored or enhanced to reduce peak flows downstream.

Topography and other developments may preclude opportunities for reservoir or washland storage. In these cases, channel improvements and flood walls can reduce urban flooding. Channel capacity can be enlarged in a number of ways, including deepening (or raising the banks with walls), widening, straightening, or even by replacing or duplicating the river channel. Such artificial channels are an effective method of reducing flooding along specific lengths of a river, but they are not without costs. The "improved" channels often require frequent removal of sediment and they can be unsightly and of low ecological value.

Alternatively, flood barriers can be built along the riverbank through a town. Advances in engineering mean that temporary barriers are now possible, installed just before predicted floods. This avoids the need for permanent high walls that would isolate a river from its town.

In any case, the objective of channel improvement and building flood barriers is to contain the flood water and to convey it quickly downstream, away from urban areas. This

Floodwaters from the River Severn in Shrewsbury Abbey, 1947. (Abbey Church of St Peter and St Paul, Shrewsbury)

can exacerbate flooding further down the river and is a highly significant issue for major rivers, such as the Severn and Thames in England, that pass through several cities, and even more so for large rivers, such as the Rhine in mainland Europe. Modern schemes try hard to avoid or overcome these problems, but where space is restricted in urban areas, compromises may have to be made. A vision of large rivers as "strings of beads", with channelised reaches through urban areas separated by large beads of flood-plain washes, managed as an integrated system, is gaining support along some major rivers, such as the Mississippi in America.

BOX 3.17 **The bittern**

A shy, secretive bird, which is more often heard than seen, the Bittern can be found in the reedbeds of wetlands and marshes. (www.rspb-images.com, Chris Gomersall)

Local school children plant the first reeds in a giant reedbed at Langford Lowfields, helping to turn an old quarry site into a wetland reserve. (Newark Advertiser Co. Ltd.)

The bittern is a secretive bird of reedbeds dominated by the common reed *Phragmites australis*, where it feeds principally on fish and amphibians. It is one of the UK's most threatened species. In the late 1960s there were 70 breeding pairs found across eight counties. Since then the number of males has declined to below 20. It is now confined almost entirely to the lowland marshes of Norfolk, Suffolk and Lancashire. The decline has been attributed to a loss of suitable large reedbeds through drainage and water abstraction, water pollution by pesticides and heavy metals, saltwater intrusion into coastal reedbeds and the decline in food availability, especially eels. Traditionally, bittern habitat was managed to produce thatching reed, but this practice was in decline by the 1930s.

The bittern is protected under the EC Birds Directive and the Bern Convention. It has been included in the UK Government's Biodiversity Action Plans. A grant of £1.5 million was given by the EU's LIFE Programme in 1999 to restore and extend bittern habitat across the UK. Scientific research has shown that the optimal size of reedbeds should be at least 20 ha (nearly 50 acres). In a series of ambitious projects, reedbeds are being restored and new ones created to provide suitable habitat for a population of not less than 100 males by 2020.

One project involves the conversion of sand and gravel quarry workings along the lower Trent to create 40 ha (nearly 100 acres) of continuous reedbeds providing a home for three or four young pairs. The reserve will be open to the public and there will be a visitor centre. Local schoolchildren are already involved in helping to plant the first 10 000 reeds as part of their studies on the environment, habitats and conservation.

Clean urban rivers can be valuable for recreation. The River Nene in Peterborough, for example, is popular for angling and is also used for canoeing and pleasure boating. This contrasts strongly with the River Tame in Birmingham, which has little recreational value.

Aesthetics, amenity and nature conservation

Rivers and streams have a special attraction for people, and the value of watercourses as a basis for planning open space has long been recognised. Water quality is important for maintaining fish stocks for angling, and for minimising health risks for contact water sports. It also has general amenity value for bankside recreation if there is an absence of odours, litter or algal scums. Visual pollutants, including recognisable sewage-derived objects and biological pollutants (for example methane bubbles rising, sewage fungus and excessive algal growth), are a source of concern and result in complaints from the public. As the value of water in the urban environment is increasingly being recognised in attracting economic regeneration, pollution issues are becoming more significant. Rivers provide great opportunities for use as amenity and educational resources but for maximum value they must be clean enough to allow safe access.

At the beginning of the 20th century, best practice in urban drainage advocated the covering of all drains, putting many streams and small rivers underground. Although culverting remained widespread in the second half of the century, many streams were impacted by other practices. Because the expansion of impermeable surfaces increased runoff during storms, many streams were canalised to accelerate the passage of floodwaters through the urban area.

Obstacles such as meanders and islands, which slow down the flow and increase flood levels, were vigorously removed. Channels were engineered to a smooth, straight form with trapezoidal cross-section to maximise the rate of water flow. To prevent erosion, channels were often lined by concrete. Bankside vegetation was removed to further accelerate flood runoff. Put simply, everything was done to get water off the land and down to the sea as fast as possible.

This policy created the problem of the "backyard" river where river frontages became neglected and used as dumping grounds for wastes and fly-tipping. It significantly reduced the landscape and wildlife value of rivers, and merely transferred the flood problem downstream.

Wildlife that evolved to exploit the diversity of habitats created by natural channels has been badly affected. Numbers of water voles have declined and bird habitat, for example for reed warblers, has been lost. The removal of gravel bars has eliminated important habitat for many insects. Drainage of flood-plain wetlands has eliminated habitat for snipe, lapwing and redshank, and caused the loss of a wide range of wetland plants.

Returning river corridors to nature

River rehabilitation can be undertaken to varying degrees. Full restoration to a predevelopment condition is not feasible in urban settings, but much can be done to enhance the environmental quality and in some instances to "naturalise" a reach of river. Improvements in water quality, regulating rapid runoff and controls on litter, are all important elements of urban river restoration. Moving the river from the backyard to the front garden, to become the focus for urban revival, requires the education and involvement of all members of the urban community.

Black Brook, Loughborough. This type of channel maximises the quantity of water that can flow during storms, but is unsightly and has little potential to support wildlife.

River rehabilitation involves the integration of actions to enhance the ecological potential and cultural heritage of stream corridors. Indeed, many of the works on rivers, including mills and water pumping stations, are of considerable interest in their own right. Bazalgette's sewerage scheme for London, built in the 1860s, included the Abbey Mills pumping station, which has been acclaimed as one of the most magnificent examples of Venetian Gothic

Abbey Mills pumping station (Abbey Lane, London E15), part of Bazalgette's sewerage scheme, is a magnificent example of industrial architecture.

Once a narrow straight channel with little ecological value, the River Cole in Birmingham has been converted to a wider channel with a range of water depths, offering more habitats for wildlife.

industrial architecture in Britain. It also included the building of the Victoria, Albert and Chelsea embankments to house London's new sewers, and the city's first underground railway line (now the Circle and District).

Project Kingfisher, River Cole

The River Cole flows through a highly urbanised catchment southeast of Birmingham. The river is a combination of straightened and meandering reaches, mostly 5–10 m wide with steep banks. In the past the river was heavily modified to improve flood protection, with "hard" engineering structures such as sheet piling and concrete used to restrain it in some places. These works degraded the wildlife, landscape and amenity value of the river. Urban runoff caused further problems.

In the early 1980s the local councils, Severn Trent Water Authority and local volunteer groups embarked on an ambitious scheme to improve habitats and the landscape along 7 km of the River Cole.

Project Kingfisher was launched in 1985 and has since seen the removal of engineering structures and the construction of pools and wetlands. Some of the major works included the removal of sheet piling reinforcement from 500 m of riverbank, and the removal of 150 m of concrete bank reinforcement and 20 m of the concrete channel at Yardley Brook. The river has then been allowed to erode to create a

Many old buildings along rivers and streams have considerable cultural heritage and can be incorporated into restoration schemes. River Avon at Bidford on Avon, looking East; and the East Mill, Belper

natural channel. Other works include the creation of a chain of wetlands and pools in the flood plain. There are now diverse plant and animal communities along the Cole Valley. The provision of walkways and a set of stepping stones enable the community to see and enjoy the habitats that have been created.

An important aspect of the project is the fact that the ranger and the local community have worked hard to preserve and enhance the habitats after the major works have been completed. Community clean-up days, vegetation plantings and pond work are all key ingredients in the ongoing success of Project Kingfisher.

Sustaining biodiversity

National Nature Reserves, Sites of Special Scientific Interest and reserves of the Royal Society for the Protection of Birds, the Wildlife Trusts, and the National Trust, offer protected sites for many species. The UK was among 150 signatories to the Convention on Biological Diversity (Biodiversity) that stemmed from the "Earth Summit" in Rio de Janeiro in 1992. Essentially, the Convention is a commitment to conserving and sustaining the variety of life on earth.

The response of the UK Government to the Convention was to advance biodiversity action plans (BAPs), with a broad strategy for conserving native species and habitats over the following 20 years. Specifically, the BAP identifies the work necessary to improve the adverse status of priority species.

A key component of the strategy was the promotion of local biodiversity action plans. The City of York's biodiversity action plan, for example, identified wet grassland, fens and swamps as a priority habitat. Within the city, the important Wheldrake Ings comprises a set of semi-natural flood-plain habitats, with a complex and extensive mixture of wet grassland, fen, swamp and flood communities, including annual plants such as bur-marigolds and yellow-cresses. Scarce plants include elongated sedge, marsh fern and marsh stitchwort. Priority animal species include otter, white-clawed crayfish, river lamprey and salmon, which require protection under the EC Habitats and Species Directive (1992). Other species of concern include bittern, marsh harrier, snipe, water rail, reed bunting, water vole, and a number of invertebrates including beetles and snails. Activities proposed by the action plan include the restoration of 10 hectares (nearly 25 acres) of wet grassland within the city by 2010 and re-establishment of breeding populations of snipe, lapwing and redshank.

BOX 3.18 **Aliens invade**

The extensive network of river corridors makes them highly vulnerable to colonisation by invasive alien species. Some, such as Japanese knotweed (*Fallopia japonica* var. *japonica*) and Himalayan balsam (*Impatiens glandulifera*) have adapted particularly well to the difficult and often polluted urban environment and have come to dominate river banks in urban areas. Japanese knotweed is a perennial plant that forms dense thickets which can displace native plants. Himalayan balsam can also displace native plants, spreading by seeds explosively propelled from ripened pods. Another invasive species, giant hogweed (*Heracleum mantegazzanium*) poses a public health hazard because its sap causes a painful skin rash on contact. These plants all die down in winter, leaving river banks unvegetated and liable to become unstable and prone to erosion.

It is not just exotic plants that pose a threat for native species. Non-native animals, such as mink and North American crayfish farmed in the UK, have escaped into the wild or have been deliberately released and now compete with native species for food, space and other natural resources. The white-clawed crayfish (*Austropotamobius pallipes*) is the only native species of freshwater crayfish in the UK. It is now under threat from alien crayfish species, most notably the American signal crayfish, which causes damage to river habitats, spreads a fungal disease deadly to the native species and out-competes the white-clawed crayfish for food.

Alien species, such as Japanese knotweed and Himalayan balsam, are frequently found on river banks in urban areas.

BOX 3.19 **Sewage treatment works**

Although sewage treatment works were not designed to attract wildlife, they often support a rich bird life. Generally, the value of such sites for wildfowl depends upon their size and the degree of openness. It also depends on the degree to which areas are used in rotation, ensuring that some areas are always in a suitable condition for feeding birds and others are undisturbed by human activity. If sites become overgrown and more enclosed their value declines, so that to maximise their value for wildlife, continuous maintenance is required.

Minworth sewage treatment works is the largest inland sewage works in the UK, occupying over 270 ha of land and treating an average of 550 million litres of raw sewage every day from the West Midlands conurbation. About 55 hectares of the site are taken up with sludge drying beds which provide habitat for green sandpiper and snipe. A series of lagoons is maintained to store excess sludge, and the rich invertebrate food source associated with the organic matter attracts good numbers of teal, shoveler and mallard ducks. Some of the dried sludge is dumped on an on-site tip with a lake surrounded by dense stands of common reed, and despite the dubious quality of the water, the lake attracts large numbers of waterfowl, including breeding little grebe, tufted duck and shoveler.

The Wick St Lawrence sewage treatment works at Weston-super-Mare lies adjacent to the Severn Estuary, close to important bathing beaches. In order to cope with an anticipated expansion of the conurbation and to reduce levels of infectious microbes in the discharges from the works, shallow lagoons (1.5 m deep) have been excavated to intercept the outflow from the original system. The capacity of the lagoons was designed to allow a retention time of 10 days, adequate to reduce levels of infectious microbes which are destroyed by exposure to ultraviolet radiation in sunlight. The shallows were planted with common reed, and willows and alders were sited downwind of the lagoons so as not to reduce the benefits of wind-induced mixing on water quality. Islands were left to provide loafing and nesting areas for birds. Little grebe and goldeneye were the first colonisers, shortly followed by mute swans. Other birds are expected to be attracted as the vegetation establishes itself.

Minworth sewage treatment works processes most of the sewage effluent from the West Midlands including Birmingham. This photograph shows part of the works being upgraded. The whole works measures 270 hectares – the cars in the car park give an idea of the size. The sludge drying lagoons (squares in the left middle-distance) are frequented by water birds. (Severn Trent Water plc)

The Little Grebe is the commonest native grebe and is found on many inland waters, including ponds and ditches. (www.rspb-images.com, Michael W. Richards)

Water works and nature conservation

A feature of urban areas are the treatment works processing sewage and industrial effluents so that water can be returned to streams and rivers in an environmentally acceptable form. These treatment facilities and any sludge lagoons have long been associated with wildlife. Many cities also have water storage reservoirs, originally built to provide local domestic water supplies or to support canal systems. The creation of an extensive series of large lakes in the valleys of the Thames and Lea had a dramatic impact on the bird life of London, creating one of the best centres for viewing a wide variety of winter visitors in the UK. A survey of December 1937 recorded over 6000 birds comprising thirteen species.

Old sewage farms, which simply stored untreated sewage in large lagoons, often attracted large numbers of wildfowl, especially waders. Modern water treatment systems still attract wildlife, and not just the tertiary treatment facilities with their reed bed systems, lagoons and grass plots, although these offer considerable potential.

River rehabilitation aims to enhance the environmental quality of a river.

Advancing corridor restoration

The rehabilitation of degraded rivers is of growing interest to practitioners of river management worldwide. In urban areas, this interest has been raised by the recognition of the value of river corridors for amenity, recreation and nature conservation, and the role of river corridors in enhancing the quality of urban living and in connecting different parts of the landscape. Today, the approach is to hold back flood waters, to restore flood plains and to create suitably designed flood and balancing areas with an interlinked green – and traffic free – network of river corridors. These corridors serve to connect different areas of the city, and sites of cultural and natural heritage.

BOX 3.20 **The River Skerne restoration**

One of the sewer outfalls, now removed.

The River Skerne in Darlington is at the heart of industrial north-east England. From the mid-19th century onwards, the natural meandering river became increasingly canalised, with a straight channel between artificially raised banks to prevent flooding in the area. In summer the banks were overgrown with alien Himalayan Balsam. With 13 sewer outfalls in less than a kilometre of river, by 1995 the Skerne had become a drain fit only for the disposal of rubbish and waste. However, since the river flows through open space in an otherwise densely urban area, it had the potential to become a valuable resource for quiet recreation.

The ecological interest of the river and its perceived environmental value were increased by restoring a meandering channel, complete with quiet backwaters essential for young fish and many aquatic insects. The original meanders could not be restored because of the need to avoid underground services installed during the last century. The banks were reprofiled and the level of the surrounding land lowered to restore the river flood plain, simultaneously making the water safer and more accessible. These actions also preserved the ability of the channel to convey flood flows and provided an element of downstream flood alleviation, produced by water storage in the flood plain.

Although the thirteen drain outfalls nominally discharged clean surface water, it was clear that wrong connections to the drains were leading to the discharge of oil, detergents and raw sewage into the river. Additionally, the standard drain outfalls were unsightly and a possible safety hazard when surrounded by high summer vegetation. The wrong connections which polluted the drains were identified and rectified. The outfalls themselves were replaced with others of a new design, and these discharge underwater and are visually unobtrusive.

The river now provides a valued resource for a variety of recreational activities.

The restored River Skerne now meanders through attractive open space.
(All images from the Northumbrian Water Group)

As a result of the restoration project, a visually unattractive channel of low ecological value has been transformed into an attractive leisure feature of growing ecological interest. Usage of the parkland has increased greatly.

Community participation, managed by a full-time liaison officer, was crucial to the scheme, with 82 per cent of the population finally in favour. Participation went beyond consultation, with schools and other community groups being involved in planting and other work. This has encouraged ownership of and pride in what has been achieved.

The project was managed by River Restoration Project, a private non-profit organisation, along with Darlington Borough Council, the Environment Agency and Northumbrian Water Ltd. Further support was provided by English Nature and the Countryside Commission, with funding from the EU LIFE fund and the Heritage Lottery Fund.

Urban water systems

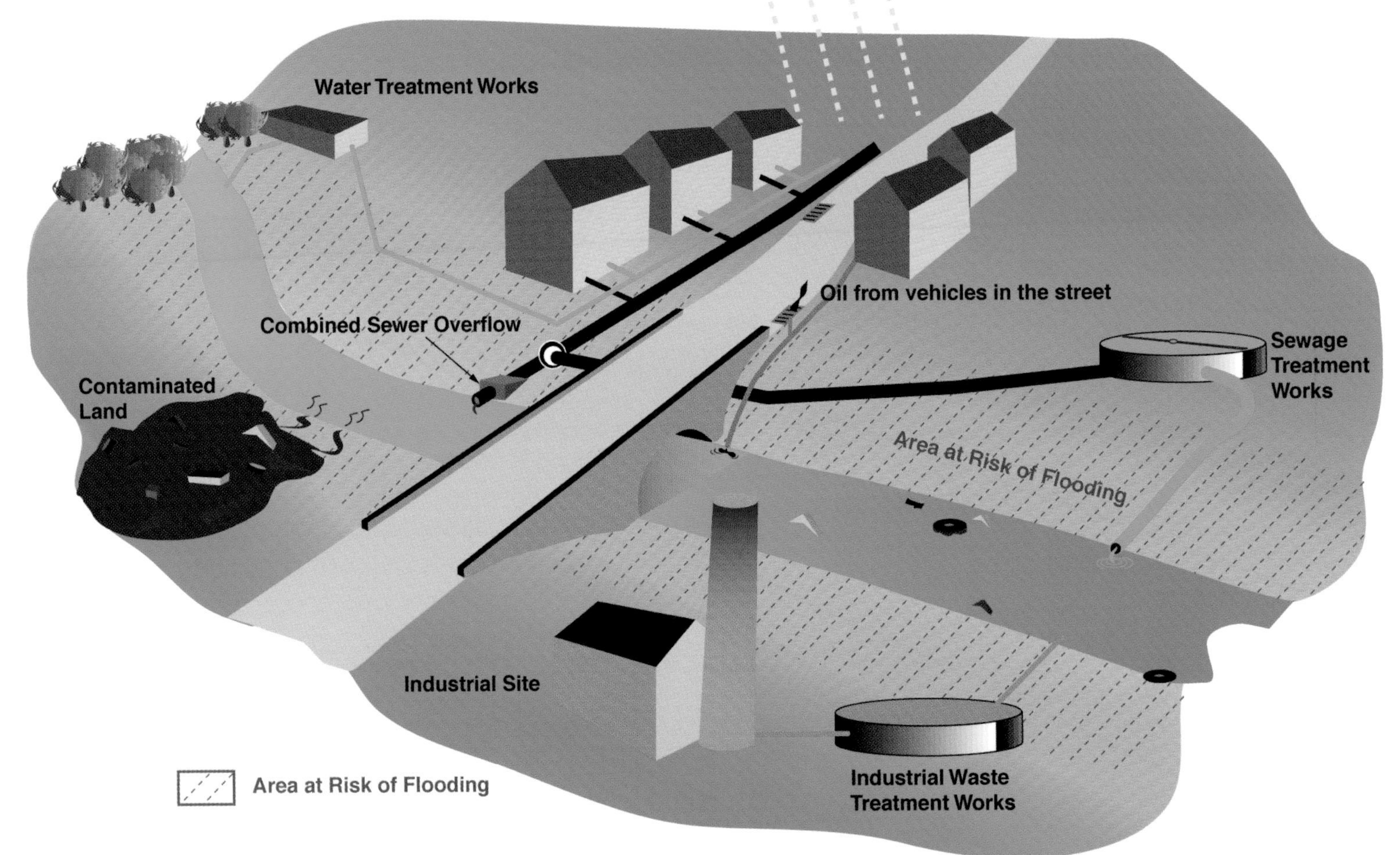

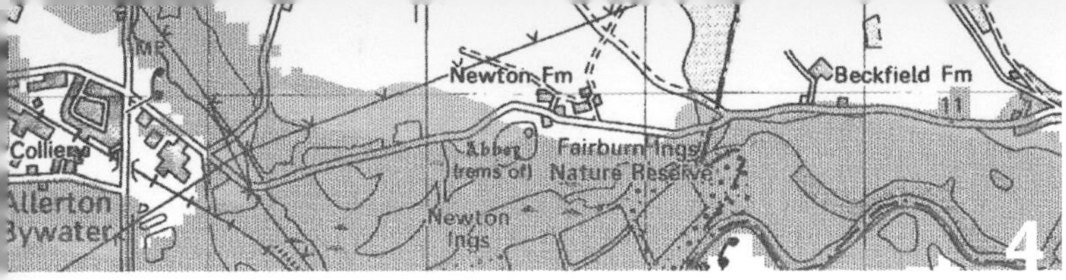

4 Sustainable urban rivers

Urban living in towns and cities can be revived around the regeneration of urban streams, rivers, flood plains and waterfronts. The approach is inclusive, comprehensive and long-term. It is also sustainable. There are two strands to achieving the vision of sustainable urban rivers: planning land use to make best use of the river, and planning the river itself, so that it is seen as an asset to the urban community.

The examples presented here show how a combination of carefully planned strategy and strong co-ordination and administration works can result in the regeneration of derelict riverside land, with the river corridor becoming a catalyst for the transformation of the wider urban area. Public participation is an important part of the planning and management process and is supported by the land-use planning framework and appropriate regulations.

PLANNING FOR THE ENVIRONMENT

The UK's land-use planning system has been operating to the same general principles since 1947. It consists of two key elements: development planning and development control. Together these two processes have been important in ensuring that impacts on the environment by development are minimised and in securing improvements in the quality of life, particularly at the local level. Since the early 1990s there has been a significant shift towards viewing the planning system as a means of achieving sustainable development.

> BOX 4.1 **Sustainable development**
>
> Sustainable development is that which meets the current needs of this generation without compromising those of future generations. It is about leaving for our children a world that we would have been happy to inherit. Sustainability requires recognition of the value of clean air, clean earth and clean water; of wildlife and of natural ecosystems; and involves a responsibility to invest in remediation, mitigation and restoration measures when contamination or ecological damage does occur. It is as much about improving our quality of life as it is about environmental protection.

The land-use planning system within local authorities has an important preventative role and is also among the most open and responsive elements of control in terms of inputs from interested parties and the wider public. It is this characteristic of planning that has important implications for the development of sustainable strategies for managing and enhancing our urban rivers.

The land-use planning system has traditionally dealt with spatial issues. It has relied on constraints that can be readily shown on a map as its basis for environmental protection, and incorporating scientific and qualitative concerns has therefore not always been easy. Increased public pressure to address a wide range of environmental issues is now resulting in an integration of urban physical planning and environmental management.

In addition to the land-use plans, there are other important plans which can contribute to sustainable land use and which interact with rivers. These include plans for protected areas, such as national nature reserves and special areas of conservation (drawn up by relevant conservation agencies), national park management plans (drawn up by national park authorities) and biodiversity action plans, which focus resources to conserve and enhance biodiversity by means of local partnerships between local authorities, conservation bodies and local communities. The Environment Agency in England and Wales also draws up non-statutory Local Environment Agency Plans (LEAPs). Public consultation is important in all these planning processes.

RIVER FLOOD PLAINS

Special consideration is now given to flood plains in planning urban development, and to this end up-to-date and consistent maps of flood plains have been produced. The Environment Agency produces these maps for England and Wales and they are available on their website alongside other river-related environmental data.

Flooding is a natural event and the storage of water on flood plains is important in controlling river flows. Along natural rivers, flooding is a valuable ecological asset, for example, increasing biodiversity and fish production, and helping to sustain a balanced ecosystem. Thus, inundation of flood plains is desirable where it can occur without risk to human life and property. By adopting a policy of sustainable flood plains, the environment agencies aim to secure and, where necessary, restore the effectiveness of flood plains for flood defence and environmental purposes.

The environment agencies, through consultation built into the planning process, advise local authorities on using their powers to ensure that watercourses and flood plains can convey and store floodwater while retaining their agricultural, amenity, recreational and wildlife habitat values. It is important to restrict development that would unacceptably increase the risk of flooding to others, either directly or by reducing flood plain storage, and to guide development away from areas liable to flood or that would prejudice existing flood defences. The agencies also aim to retain and, where necessary, restore the effective flood-flow conveyance and flood-water storage capacities of flood plains and to persuade planning authorities to recognise the importance of the continuity of flood plains and river corridors, especially through settlements, for flood defence and environmental purposes.

Trying to prevent floods is an intervention in a natural process. It costs money to build and maintain flood defence works. Preserving the flood plain is an alternative and brings its own benefits. The Trent Flood Plain Initiative, for example, aims to restore and enhance some of the wetland wildlife habitats, landscapes and features of the Trent flood plain, many of which have been lost to built development and changes in land management in the catchment. Working in partnership with conservation bodies, landowners, business, industry and the local community, the Initiative hopes to advance a common strategy to benefit both people and wildlife. It seeks to build on the benefits of flood plain wetlands that offer not only value to wildlife but can fulfil a range of functions including flood alleviation, pollution control, recreation and landscape improvement.

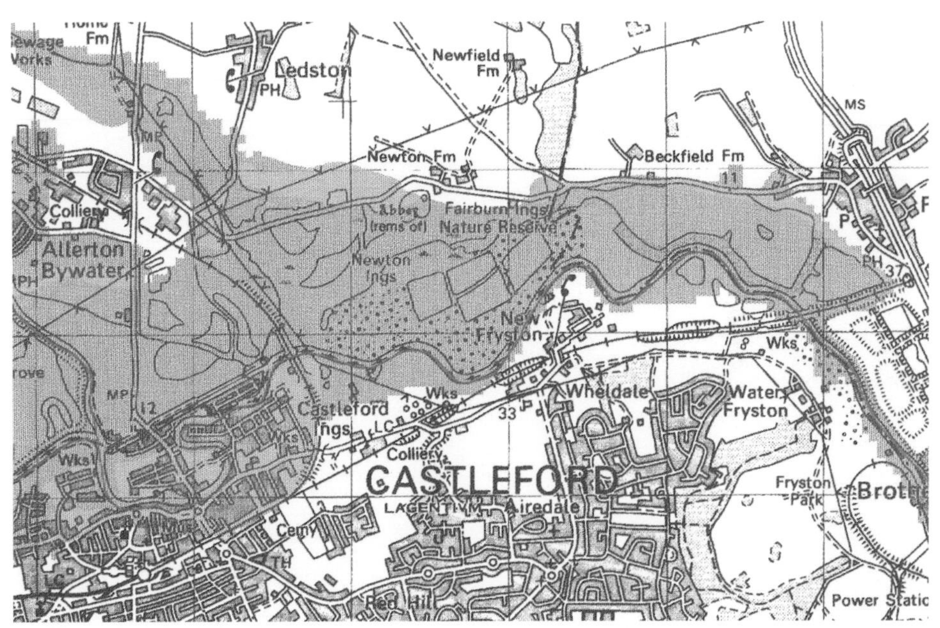

The area potentially flooded by the River Calder at Castleford includes housing as well as farmland. (© Crown copyright. All rights reserved. Environment Agency GD03177G, 2001)

BOX 4.2 **Local authority planning**

The planning process, and the role of local authorities within it, is currently defined by the Town and Country Planning Act 1990. There are two aspects of local authority planning: development planning, where a local authority actively sets out the types of developments that are needed in different areas and which may be permitted, and development control, which ensures that individual developments proposed by applicants are appropriate.

Development planning

Development planning involves local authorities preparing plans which will shape development in their areas. Structure Plans provide a policy framework, usually on the scale of a county, for land-use, environmental conservation and development control over a 15-year period. Local Plans, usually on the scale of a town, contain more detailed proposals, including the identification of areas for growth or conservation, and for industrial development. Specific plans are also produced in relation to extraction of minerals, economic development and management of waste. Development plans are subject to public consultation at their draft stage. Plans may be subject to a public inquiry and the Secretary of State can "call in" a plan for his review.

Planning Policy Guidance Notes produced by the Department of the Environment, Transport and the Regions (now DLTR) specify that plans must include policies relating to the improvement of the physical environment, the management of traffic, the amenity of land and conservation of both finite and renewable resources such as water. There is support for the environmental appraisal of plans and acceptance that a precautionary approach will need to be taken where development might have potentially irreversible environmental impacts.

The development plans are informed by Regional Planning Guidance (RPG) which takes an even longer time perspective, covering periods up to about 20 years, and provides a broad framework for the preparation of both Structure and Local Plans. RPGs are being strengthened in relation to sustainable development objectives and will, in future, include regional transport strategies and provide a framework for the strategies of Regional Development Agencies.

RPG may, for example, encourage the active conservation of water by recognising potential water supply constraints on new development and attempting to steer it away from areas where there may be potential shortfalls in supply. It may also recognise water quality as a factor in considering future development allocations in consultation with the water companies. The need for investment in sewage treatment and discharge systems may be highlighted so that water companies can plan to meet effluent and quality criteria where new development is being considered.

Development control

Local authorities have a duty to control development, in response to applications. Development control requires most types of development, including changes of use of land, to have a planning permission. Any development control decision taken by the local authority must be in accord with the Structure and Local Plans, unless there are overriding reasons for departing from these policies. Thus these plans are important in defining what is appropriate development, and their conservation and other protection policies are important in ensuring that inappropriate development is not permitted. Once an application for planning permission is received the local authority has a fixed time in which to consider the application. During this period it will consult with statutory bodies, such as the environment agencies and nature conservation bodies, and notification of the application will also be sent to neighbours of the site. Other interested parties locally, such as conservation groups, may also be consulted. The planning officer for the case prepares a report for the local authority Planning Committee which has to take into account any comments received as a result of this consultation process. The Committee will either approve, conditionally approve or refuse the application. The applicant has the right of appeal against the decision to the Secretary of State. An appeal is considered by a planning inspector appointed by the Secretary of State and in certain cases the appeal may be heard at a public inquiry.

For developments which may have a significant environmental impact (as defined by the Town and Country Planning (Environmental Impact Assessment Regulations) 1999, an Environmental Assessment is required and the applicant has to submit an Environmental Statement with the planning application. The Environmental Assessment process is an important means of understanding the impacts of a proposed development on human beings, flora and fauna, water, soil, air, climate, landscape, and the material assets of a local area. It is also a means of ensuring that appropriate measures to reduce and mitigate potential impacts are designed into the project at an early stage. Examples of projects which could have significant impacts on the river environment include large residential developments, power stations, road schemes, large industrial developments, as well as projects which are designed to impact on rivers such as flood diversion channels, dredging and sewage treatment works.

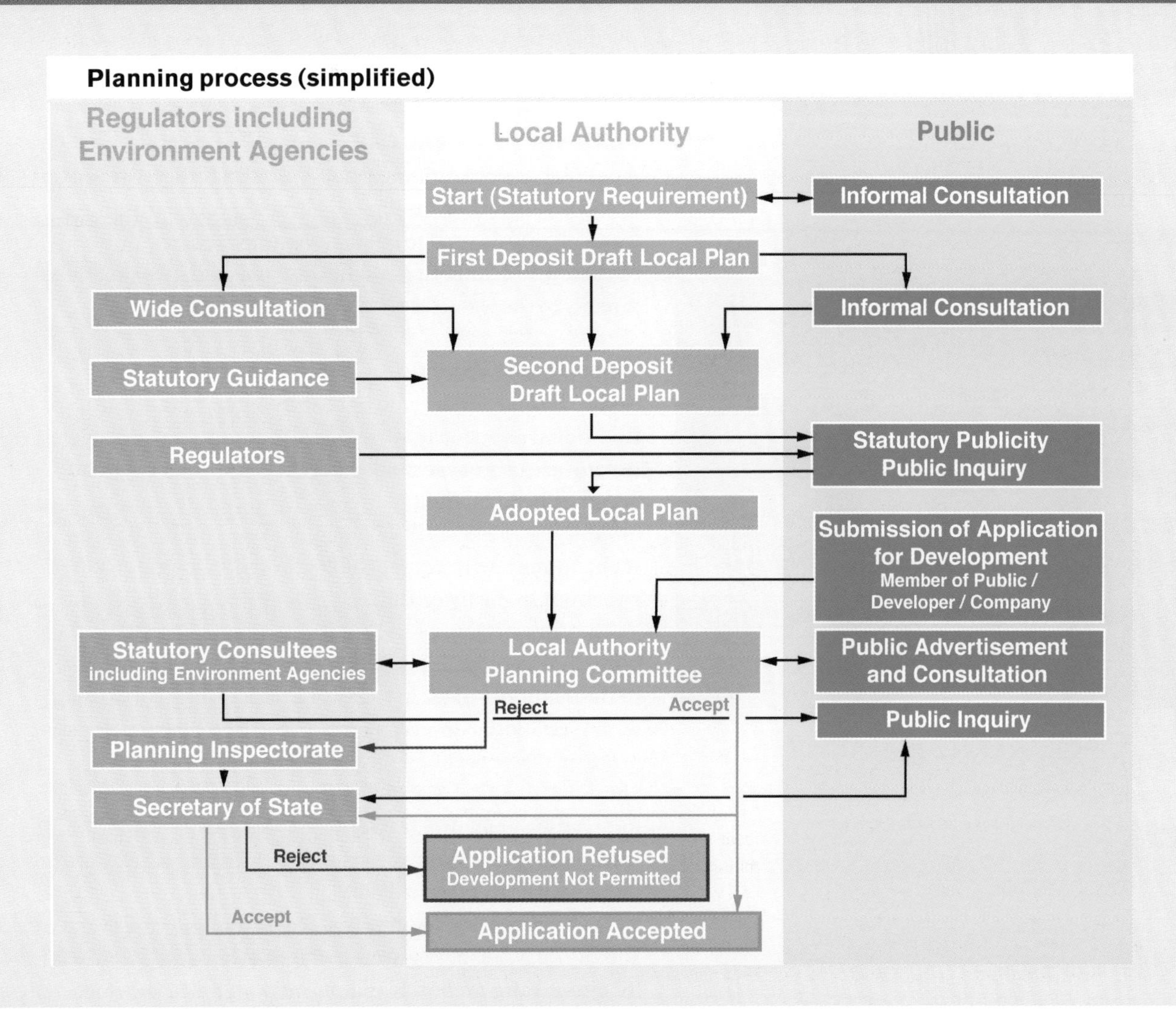
Planning process (simplified)
Regulators including Environment Agencies
Local Authority
Public
Start (Statutory Requirement)
Informal Consultation
First Deposit Draft Local Plan
Wide Consultation
Informal Consultation
Statutory Guidance
Second Deposit Draft Local Plan
Regulators
Statutory Publicity Public Inquiry
Adopted Local Plan
Submission of Application for Development
Member of Public / Developer / Company
Statutory Consultees including Environment Agencies
Local Authority Planning Committee
Public Advertisement and Consultation
Reject
Accept
Public Inquiry
Planning Inspectorate
Secretary of State
Reject
Application Refused
Development Not Permitted
Accept
Application Accepted

The elevated roadway testifies to a history of flooding from the River Trent. The town of Burton-on-Trent (in the distance) has been built away from the river flood plain.

PLANNING IN PRACTICE

A polluted watercourse may not be an absolute constraint on urban regeneration, but incremental improvements in river water quality do tend to open up significant new opportunities for both development and activities that relate directly to the waterside.

Tyneside and Wearside

Newcastle and Sunderland both suffered from the decline of traditional riverside industry in the late 20th century, leaving a legacy of vacant and derelict land. The rivers had become black, smelly and lifeless. The quayside was a place to avoid. Tyne and Wear Development Corporation (TWDC) was set up in 1987, with responsibility for redeveloping 42 km of riverfront in the two cities.

The Corporation integrated the development of new industrial, retail and residential property with improvement in the quality of water in the rivers. Importantly, 25 per cent of the residential development comprised social housing, providing low-cost homes for rent and equity sharing. Substantial capital was invested by Northumbrian Water to improve water quality in the River Tyne, providing new intercepting sewers on both banks of the river, which were topped with riverside walkways.

The intercepting sewers provided a dramatic improvement and the quayside has become a sought-after location both for living, and also for hotels and the busy nightlife for which

Box 4.3 **Holme Pierrepont National Water Sports Centre**

Britain's first purpose-built, multi-water sports centre, opened in 1972, is set in a 110 ha country park and is located only 5 km from the city of Nottingham, providing an important public amenity. Designed to be a centre-of-excellence for rowing, canoeing and water ski-ing, the Centre also offers considerable scope for power boating, water-ski racing, hydroplaning, angling and sailing. The core facility, its Regatta Lake, measures 2215 m long by 315 m wide. The canoe slalom course was built in 1996 using the head of water created by the adjacent flood control sluices. The British Canoe Union has located its headquarters alongside.

The Park incorporates a Nature Reserve created from a disused gravel pit and was established specifically for disabled people from a nearby residential home. The varied flora includes marsh marigold and southern marsh and common spotted orchids, and there are good stands of Norfolk reed. Birds include the great crested grebe, kingfisher, reed bunting, skylark and lapwing with common tern, yellow wagtail, reed warbler and sedge warbler being regular summer visitors. Local schools use the park to study natural history and countryside management.

On the opposite side of the River Trent lies Colwick Park, managed by Nottingham City Council. The park includes wetland areas providing facilities for trout fishing, coarse fishing, model boats and board sailing, and has a 250 berth marina. The park is well known for wildlife, especially over-wintering birds. Both sites show how imaginative restoration of mineral workings can provide extensive wildlife habitat alongside a wide range of recreational activities.

The Holme Pierrepont watersports centre, beside the River Trent just outside Nottingham, incorporates excellent facilities for watersports with a nature reserve. (National Watersports Centre)

The marsh marigold thrives in wet marshland. (Peter Stiles Photography)

The Millennium Footbridge across the River Tyne, linking the quaysides of Newcastle and Gateshead, was installed in one piece by a giant floating crane. (North News and Pictures)

Newcastle has become famous. On the Gateshead side a major arts centre is emerging. The conversion of the Baltic Flour Mill to a contemporary art gallery is due to open in 2002 to be followed by the Music Centre, an international concert hall of the highest quality. The stunning new Millennium Footbridge connects the quaysides on either bank. None of this would have been possible without the fundamental improvement in the water quality of the River Tyne.

Crucially, the TWDC actively involved the community in the programme through education, consultation and direct participation. Cultural heritage projects were combined with riverside public art, and a water sports strategy was formulated in association with the National Rivers Authority (one of the predecessor organisations of the Environment Agency), the Sports Council, and five local authorities. The role of the urban river was obviously a pivotal one: an attraction for potential investors and a focal point for community life.

The Medway riverside

The planning system in the UK is predominantly reactive, but the River Medway in Kent provides an example of a local authority taking the initiative. The Medway riverside between Rochester and Chatham was characteristic of urban riverfront land in decline, blighted by dereliction, contamination and poor accessibility, following the rundown of the naval dockyards. Some of the land was liable to flooding. In September 1998, the council bought 32 hectares of low-value land, occupied by derelict railway sidings, builders' merchant yards and gas holders, and previously owned by 30 different parties, creating a single site for redevelopment. This was done using a Compulsory Purchase Order, at the time unprecedented in its scale outside Urban Development Corporation areas. Local authority powers were critical here; private enterprise would

Derelict warehousing along the waterfront at Rochester is earmarked for redevelopment. (Medway Council)

almost certainly not have been able to achieve this co-ordinated result.

Compulsory Purchase Orders require a considerable investment of resources, not least to compensate landowners, and their use is discouraged by the Department of Transport, Local Government and the Regions except where there is a compelling case in the public interest. The Rochester Riverside project, has been praised by Lord Rogers of the Urban Task Force as epitomising the Urban Renaissance ideal, demonstrating the potential advantages of the system in aiding urban regeneration.

PUBLIC PARTICIPATION

It is increasingly recognised that everyone has a part to play in protecting the environment. Principle 10 of the Rio Declaration on Environment and Development emphasised the role of all citizens in the pursuit of sustainable development. The ways in which local groups can participate in environment and development decisions and the need for easy access to environmental information to expand public involvement is highlighted in Agenda 21, an outcome of the Rio conference.

Public consultation and participation is increasingly important to the environment agencies. They currently

BOX 4.4 Edinburgh's key asset: the Water of Leith

The tidal mouth of the Leith provided a haven for ships and led to the growth of an important port and shipbuilding and repair industry. The power provided by the rapid flow of the river was harnessed to drive water mills, establishing a string of industrial villages. Pollution was the cost but, as the mills closed, industry declined and interceptor sewers were constructed, the river was able to face a new future. Today, the Water of Leith is an important trout stream attracting some 1000 anglers each year. The river provides "a ribbon of green" and is the single most significant natural corridor through Edinburgh, where the steep wooded banks of the river are a fitting backdrop to the old mill buildings and renovated warehouses. It also provides a sanctuary for wildlife and people, an outdoor classroom for adults and children, and a network of walkways linking and enhancing the appeal of residential areas. The river is designated as an Urban Wildlife Site by Edinburgh District Council in the Urban Nature Conservation Strategy for Edinburgh, and is a Scottish Wildlife Trust Wildlife Site.

The Water of Leith, Edinburgh, a green ribbon through the urban area. (The Water of Leith Conservation Trust)

In 1989, residents concerned about Edinburgh's river and its future established The Water of Leith Conservation Trust. The Trust's role includes:

- raising awareness of the river as Edinburgh's key asset;
- providing an educational and interpretation service;
- organising an annual river clean-up;
- setting up a voluntary ranger service to ensure the future stewardship of the river.

The Trust is in a working partnership for the integrated management of the Water of Leith led by The City of Edinburgh Council, East of Scotland Water, Forth Ports Plc, the Honorary Bailiffs of the Water of Leith, Scottish Environment Protection Agency, Scottish Natural Heritage, Scottish Wildlife Trust and West Lothian Council.

In promoting an Integrated Environmental Action Plan for the Water of Leith, the Scottish Wildlife Trust emphasised the role of community groups in ensuring that the plan is implemented in their area and the scope for community groups to "adopt" areas along the Water of Leith, to manage these areas as community wildlife sites and to implement the Plan in these areas.

All sorts of interesting "creepy-crawlies" are to be found lurking beneath the stones of a healthy river. The Water of Leith provides a valuable educational opportunity. (The Water of Leith Conservation Trust)

Sarehole Mill, Birmingham, the inspiration for J.R.R. Tolkien who lived nearby. (Geoff Dowling)

consult on a number of issues, including flood defence, licence applications, Local Environment Agency Plans and, at a more strategic level, with strategies for managing water resources. The fundamental importance of public consultation in land-use planning has already been discussed.

Community participation repeatedly emerges as an important factor in successful urban river projects. The case of Moseley Bog in Birmingham demonstrates how powerful an influence the community can be. The site centres upon Old Pool, a reserve water supply for Sarehole Mill constructed in c.1750 adjacent to the River Cole. By 1919, however, suburban Birmingham had overtaken the valley, the mill was abandoned and the pool became neglected. In 1980 a planning application was submitted for the redevelopment of the site for 22 houses, a threat that immediately galvanised the community into a dramatic response. Within 3 weeks a 12 500 signature petition was collected, 500 protest letters were written to the Council, and a public meeting attended by 300 people was held. The result was the Save Our Bog Campaign, which attracted national media coverage and widespread support, not least because the site has strong associations with the author J.R.R. Tolkien.

After 6 years of active protest, the Campaign achieved its aim. The development was reduced to 11 houses, the wetlands remained intact, and the mill was restored and opened as a Visitor Centre. The Campaign survives in the form of the Moseley Bog Local Nature Reserve Conservation

The LEAP Process

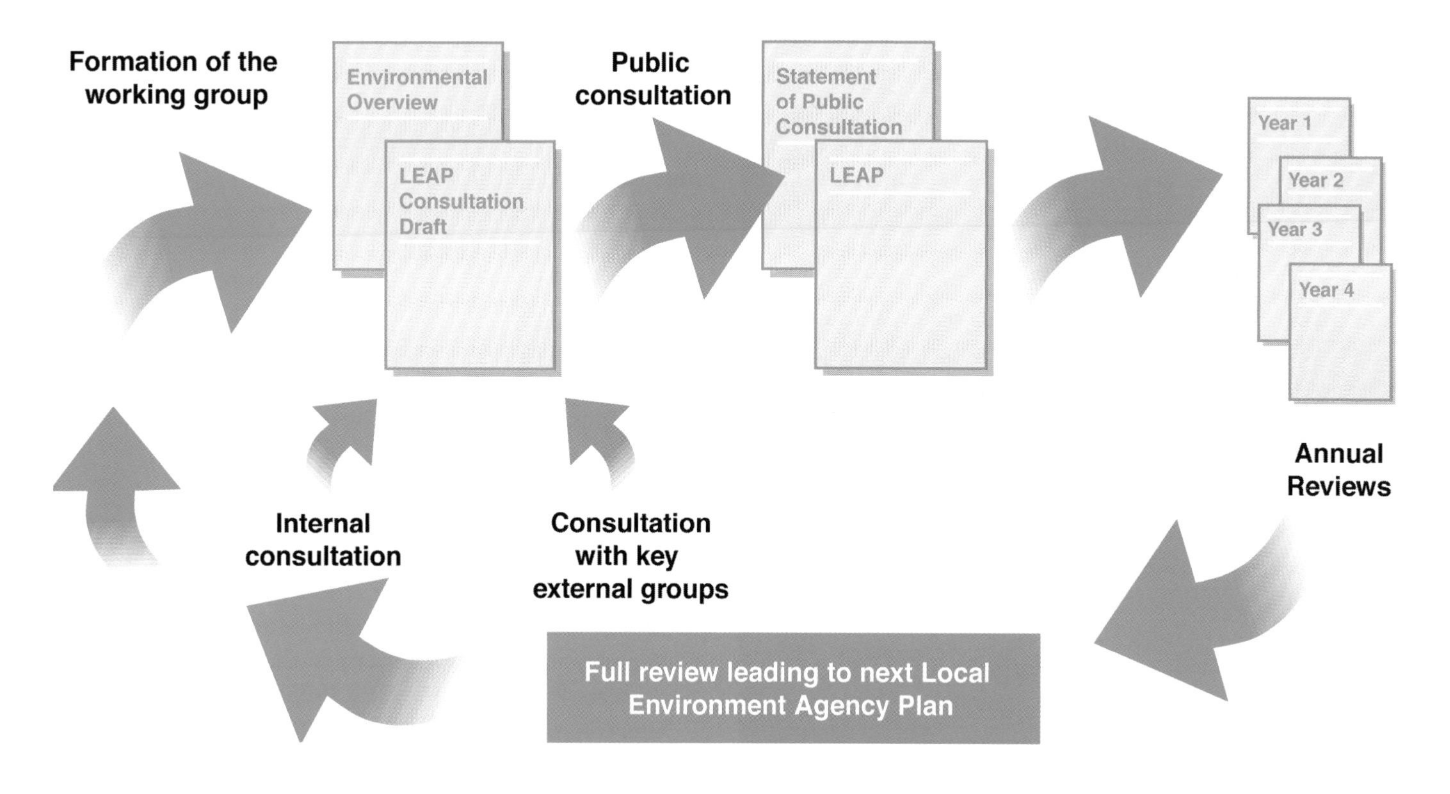

Public participation is important to the environment agencies. Consultation processes, such as LEAPs, allow involvement of all those concerned with their local environment.

Group, maintaining the link between the community and the site. Furthermore, an environmental strategy has subsequently been adopted for the length of the River Cole corridor through the city. A genuinely community-led initiative has not only secured the survival of an important urban wetland, but it has had wide implications for the urban rivers of Birmingham as a whole. It is important that, in the search for urban development sites, the value of these rural habitat fragments continues to be recognised.

The Moseley Bog case could actually be presented as a partial failure of public participation in the first instance, in that the original planning application had to force such a public reaction. At the current time many different pressures are converging for genuine public participation in decision-making, rather than mere consultation on plans and proposals and provision for access to information.

Local authorities and the environment agencies are increasingly testing new forms of participation. These seek to involve all relevant interests in the actual development of plans and proposals so as to achieve a consensus on the right solution, enhance community involvement in the design, implementation and monitoring of new development, and actively involve local knowledge in promoting sustainable development.

One of these new participation methods is 'Planning for Real', a method of involving people in decisions about their neighbourhood through building models of the area upon which agreement is reached about the right solution.

The activities of volunteer groups improve the river bank environment and build a sense of community responsibility for the river. (The Lower Lea Project)

Other methods are to use citizens' juries and community advisory groups, where small groups of people considered to be representative of a local community come together to discuss the development of appropriate plans for an area and to question expert knowledge. Citizens' juries do this over a few days and question experts to gather information so that the jury can come to a consensus view, which is presented to the decision-making authority. Community advisory groups often function over months and involve meetings, seminars and site visits.

All of these new methods of participation stress the need for the public to be involved in and to influence the decision-making process. It is expected that these methods will become even more important in the future.

CATCHMENT PLANNING

Ultimately, the responsibility for protecting our rivers, whether urban or rural, depends on the actions of everyone living in, travelling through, working or passing leisure time within their catchments. However, conflicts of interest can arise between different users: the industrial enterprise with an effluent to dispose of and the angler, for example, or the domestic consumer wanting to use a lawn sprinkler and the conservationist concerned about the effects of low flows on wildlife.

It is clear that, when pressures on rivers develop to the stage where these conflicts become manifest, a catchment-based system of planning and control is necessary. The European Union's new Water Framework Directive requires all member states to adopt Catchment Management Planning.

In England and Wales, this approach has previously been implemented through a series of Catchment Management Plans and latterly Local Environment Agency Plans (LEAPs), covering each catchment or sub-catchment. This has been driven by practical needs, with many catchments experiencing real conflicts of interest between abstraction and the need for maintenance of river flows to protect wildlife and fisheries, and treatment of effluent or dilution as a means of avoiding pollution. However, in Scotland, where water resource drivers are much less forceful because of lower population density and a wetter climate, the process is less advanced.

Within the catchment management framework, detailed control mechanisms and agreed standards are required to regulate specific aspects of river use and protection. Regulation and management can be implemented at different stages in the catchment water cycle and optimum effectiveness may be achieved by control measures a long way "upstream" from the point of environmental impact. Point and diffuse pollution sources need different approaches and other measures are needed to control quantity, both through flood management and abstraction control.

SETTING STANDARDS

Regulatory authorities need to define an Environmental Quality Objective (EQO) for each river, thereby setting the Environmental Quality Standards (EQSs) that must be met. In order to ensure compliance they must then consider the loads of pollutants entering the river, the water available for dilution (which may be reduced by abstraction or supplemented by discharges upstream) and the behaviour of the pollutants in the river. Control is achieved through a series of regulatory controls on emissions to, and abstractions from, watercourses.

The drivers for water quality regulation in a particular river are defined by use-based EQOs, which include both human users and the requirements of the natural environment. Some examples are listed here.

Use	Environmental Objectives
Public amenity	No visible pollution, no odour
General ecosystem	Maintain normal ecological community for the climatic zone and physical nature of habitat
Special ecosystem	Maintain criteria for which site is recognised as being of special interest
Drinking water abstraction	Maintain water in a suitable condition to be treatable at the water treatment facilities available
Commercial fishery	No tainting, fish fit to eat, no adverse effect on population/growth, no interference with catching methods
Sport fishery	Maintain fish populations, maintain access
Bathing, diving	Aesthetic, microbiological & toxicity standards, maintain access to bathing areas

Having defined use objectives, EQSs can be defined which will produce river water quality that meets the environmental objectives. Standards relate to things that can be measured, and are therefore needed for practical comparison. Standards for many potential pollutants have been set at a European level, through European Union Directives, and are implemented by national legal systems in member states. Examples include Directives on water for drinking water abstraction, fish life and bathing, which all set standards in terms of visual, chemical or microbiological parameters. These uniform standards establish a baseline that supports fair economic competition between nations. In most countries there is a mix of these uniform standards and EQSs. The latter are always required in cases where the uniform standards do not provide a sufficient degree of protection for local circumstances. Other Directives set qualitative standards that must be interpreted in regulations. For example, the Wild Birds and Habitats Directives require maintenance of "favourable conservation status" in sites where the EQO has been defined through designation as a Special Area of Conservation or classification as a Special Protection Area.

Other EQSs have been set to cover all uses under the Dangerous Substances Directive. These apply to all waters, whatever their use, and require control, reduction and elimination of a wide range of defined toxic or harmful substances. Similarly, the Water Framework Directive will apply to all river waters and sets qualitative standards that require maintenance of "good ecological quality".

The EQS system has also been applied on a national basis in the UK for general river ecosystem protection, through the Rivers Ecosystem classification. Target river quality objectives have been set under this system for all significant watercourses in England and Wales.

BOX 4.5 Monitoring and enforcement

The environment agencies undertake routine surveillance activities, to comply with their duty to assess and report on the state of the environment. Where standards or targets are set for particular water quality parameters, for example as part of a consent to discharge effluent into a river, monitoring is carried out in order to demonstrate compliance against these targets. Monitoring is also required to show compliance with abstraction licence conditions and planning conditions and agreements. In the case of a pollution incident, a rapid monitoring response may be necessary to assess the possible impact and plan an emergency response. Monitoring will be essential to collect the evidence needed to prosecute the culprit.

Environment Agency inspectors regularly collect river water samples for quality checks.

Water quality monitoring is a complex field. Analysis of trace amounts of toxic chemicals in water can be technically very difficult and expensive. For example, access is often practically difficult, especially when trying to trace a pollutant back up a pipe to find its source; and in the case of some materials, analytical methods are not yet sensitive enough to trace the source or fate of the pollutant in the river.

Enforcement of regulations may, in the final resort, require recourse to prosecution. Magistrates' Courts can impose penalties up to £20 000 and 6 months imprisonment for water pollution; higher courts can impose unlimited fines and 2 years' imprisonment (the largest fine yet levied is £4 million, although this was reduced on appeal). The agencies have powers to enforce cessation of activities likely to cause pollution and can also recover the costs of remedial works from the perpetrator of pollution. After a successful prosecution, costs are often awarded against the defendant, and may amount to more than the fine. Civil court proceedings may also follow, with damages sought by affected parties, such as angling clubs following fish deaths or a water company whose abstracted water has been contaminated. Again, these costs often far exceed the value of fines levied.

The EQS scheme, whilst a useful measure of water quality for the purposes of the Environment Agency, does not readily translate into land use planning. However, incremental improvements to water quality do open up additional opportunities for land uses and development in general and for activities related to the waterside in particular.

A river with very poor water quality has no amenity value or recreational appeal, and developers turn their backs on it. As water quality is upgraded so that it no longer smells or presents a public health hazard, developers and the public become prepared to face the river again. Volunteer groups can be encouraged to clear rubbish and to help with habitat improvements, such as reed planting. With further improvements of water quality to fishery standard, wildlife and anglers will return and opportunities are created to develop nature-based activities, including educational programmes. The waterside location also becomes a significant selling point for both residential and commercial properties, and restaurants, bars and cafes are attracted to the waterside.

REGULATION

Plans are an essential step towards achieving a desired outcome, but there needs to be a process to ensure that plans are followed. Therefore, both local authorities and the environment agencies have regulatory functions. Regulation is achieved in a number of ways:

- **Encourage** Positively encourage design of activities to produce the least impact on the environment
- **Prevent** Evaluate planned activities and forbid those where the risk of environmental damage is unacceptably high;
- **Consent** Limit discharges so that they do not cause pollution;
- **Prosecute** Prosecute when damage occurs, and try to recover enough money to repair the environmental damage caused.

There are clear advantages for the environment in favouring the higher levels of the hierarchy and these ideas are now being captured in legislation. Modern regulations based on European Directives require a prior investigation of the pollution potential of a wide range of activities and the demonstration that suitable measures are in place to prevent the possibility of pollution, before the activity commences.

Controls on activities affecting rivers in England and Wales are exerted through regulations under the Environmental Protection Act 1990, the Water Resources Act 1991 and the Pollution Prevention and Control Act 1999 (the same ends are achieved by different legislation in Scotland and Northern Ireland). These controls are intended to ensure that the environmental standards for a particular river are complied with, and are enforced by the Environment Agency in England and Wales. The controls cover a wide range of activities that have the potential to affect the environment, including over 2000 industrial processes, 100 000 discharge

BOX 4.6 **Pollution of the Tyne**

1 Acid rain and acid mist continue to affect upland watercourses.
2 Waste from former lead mines, the last of which closed 30 years ago, has the potential to leach lead, cadmium and zinc into the River Nent. Reclamation works have now been undertaken to reduce incidents of polluted water entering the river.
3 Sheep dip leakage was a problem in Allendale in the 1980s, affecting both fish and the invertebrates on which they feed. Northumbrian Water plc now collects waste sheep dip in the Derwent catchment, which is used for public water supply.
4 Phenol contamination of the River Tyne in 1991 affected a public water supply intake, causing a foul taste in drinking water. The source of pollution was never found, although an assessment have been carried out to reduce the risk of a repeat incident.
5 Interceptor sewers have captured discharges to the river since December 2000, for treatment at Howdon. Recently upgraded, Howdon Works provides primary and secondary treatment conforming to the Urban Waste Water Treatment Directive for sewage from 1.2 million people.
6 'Bovril boats' dumped 500 000 tons of sludge from sewage treatment each year until 1998, when the practice ceased as a result of the Urban Waste Water Treatment Directive. The sludge is now transported to the regional sludge treatment centre at Bran Sands (see Box 3.11).
7 Mine water discharge to the River Team from the former Kibblesworth Colliery.
8 Kielder reservoir regulates the flow of the River Tyne and thereby supports water abstractions downstream. Drinking water for Newcastle is abstracted at Horsely, near Wylam. Water can also be pumped out of the River North Tyne to support the River Derwent, and, further to the south, the Rivers Wear and Tees.
9 & 10 Paper mills. Direct discharge into the River Tyne is regulated by trade effluent consents to protect the environment.
11 Wood chip factory. Cooling water discharge into the river is regulated by discharge consent to protect river water temperature.
12 Commercial coniferous forest, leading to peat erosion and acidification of the water.

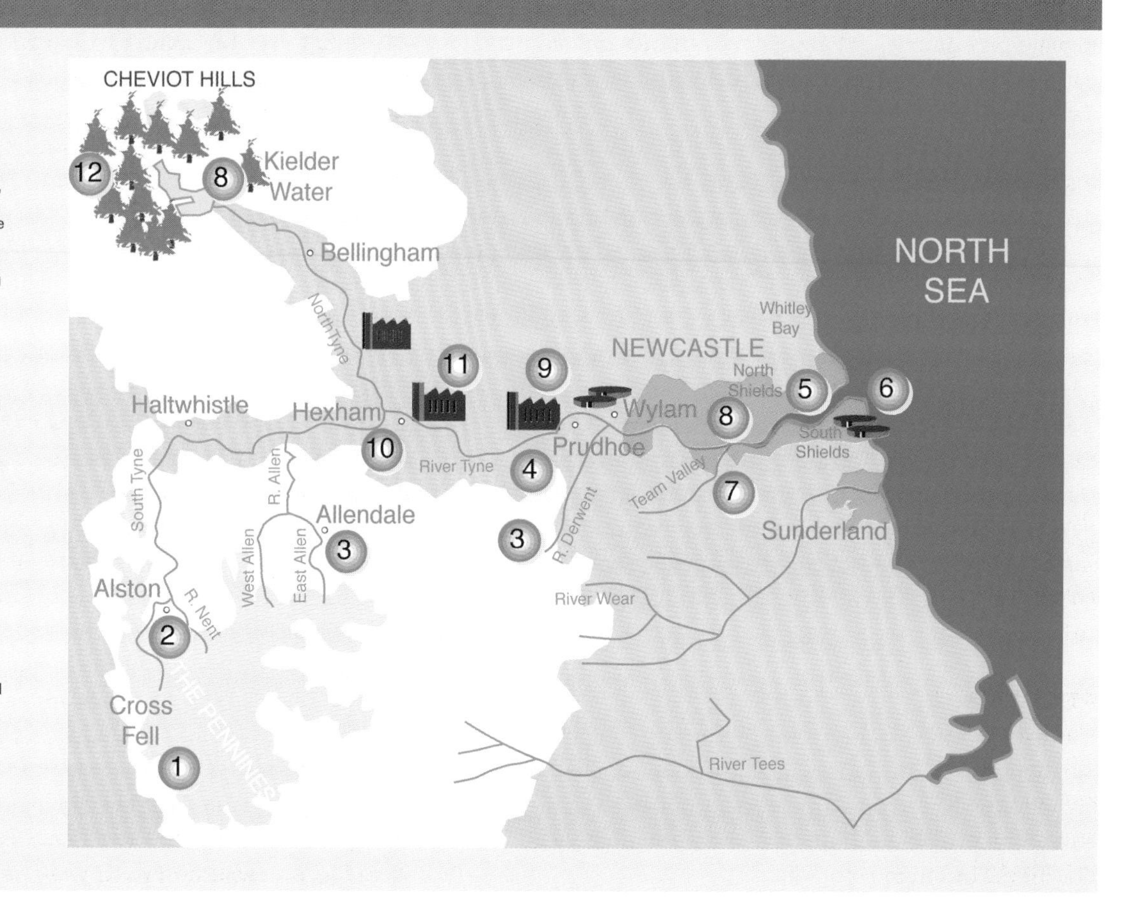

consents to water and 50 000 licensed water abstractions in England and Wales alone. The Agency therefore has a key role to play in monitoring, protection and enforcement and a considerable contribution to make to sustainable development.

Points of control

Control affecting the riverine environment can take place at various points. Control of abstractions, for example, may take place well upstream of the river reaches of concern but can affect the water available to maintain habitats or dilute downstream discharges of waste water. Other examples include control of runoff processes through influencing drainage practices, development control and flood plain management; the control of pollution sources, through regulation and advice in the case of diffuse sources and by consents and authorisations for point sources. These individual points of control must become linked in a coherent strategy if a catchment planning approach is to be successful.

REPORTING BACK TO THE PUBLIC – ACCESS TO INFORMATION

The environment agencies have always recognised the vital role which access to information plays in helping to achieve environmental goals. It is public influence that helps achieve sustainable environmental progress and therefore the environment agencies of the UK have a legal duty to report the state of the environment to the public. The agencies also actively encourage the public to seek environmental information and publicise access rights and details of the information available.

The agencies hold a wide range of information relating to the water environment, including: water quality, water resources, integrated pollution control, flood defence, fisheries, navigation, recreation and conservation. In England and Wales, reporting is already well established, in the form of the Environment Agency's General Quality Assessment (GQA) system, which has a number of separate categories of quality reporting, covering general chemical quality, biological water quality, nutrients and aesthetic quality. The data feed into national water quality reports and published State of the Environment reports, which report upon progress and trends over time and from place to place and are used in setting national policy and informing development plans. The importance of reporting is emphasised in the EU Water Framework Directive, which will require reporting of water quality and ecological quality.

BOX 4.7 **Gaining access to information**

In the UK there are a number of Regulations that guarantee access to environmental information, derived from the European Directive on the Freedom of Access to Information on the Environment (90/313/EEC). The purpose of the Regulations is set out in Article 1 of the Directive: "to ensure freedom of access to and dissemination of information on the environment held by public bodies".

Under the Regulations any individual can request environmental information from a public authority or other body with public responsibilities for the Environment. Environmental Information is that relating to the state of the environment (water, air, flora, fauna, soil or any natural site or other land), activities which adversely affect the environment and measures being taken to protect it.

The Regulations apply to information recorded in any format and would include, for example, reports and computer records. There is a duty on the public authority to respond as soon as possible, but within a limit of 2 months. This response must either contain the information requested or specify reasons why the information could not be supplied. A "reasonable charge" to recover the costs of providing the information may be made.

Public Registers are complementary to the regulations on access to information on the environment. There is a duty to hold and maintain particular sets of information and make them freely accessible for inspection. The information each register contains varies according to the particular legislation that provides for it, but can include applications, authorisations and monitoring information.

Public registers include many activities requiring an authorisation of some kind, including:

- the Integrated Pollution Control Register which includes information on discharges from major industrial sites;
- the Waste Licensing Register which includes information on all licensed waste treatment and disposal sites;
- the Water Quality and Pollution Control Register which includes information on river water quality and discharge consents, where permission has been granted to release effluent into a watercourse, and water sampling information;
- the Water Abstraction and Impounding Register which holds details of applications for abstraction licences and the agencies' decision. Public registers are available for inspection by appointment at environment agency offices.

In addition to general information on the state of the environment collected as part of routine monitoring, various regulations such as those enabling the granting of discharge consents require information to be placed on public registers. Public registers are freely available for consultation by the public at environment agency offices.

A further useful source of environmental information are reports published by both the public and private sectors. Many of these, including government policy documents on the environment, can found in public libraries. Other environmental information, such as local registers and plans, is available for public viewing at local authority offices.

The Internet is increasingly a source of information, with all the UK agencies now having sites, providing policies, reports and customised information.

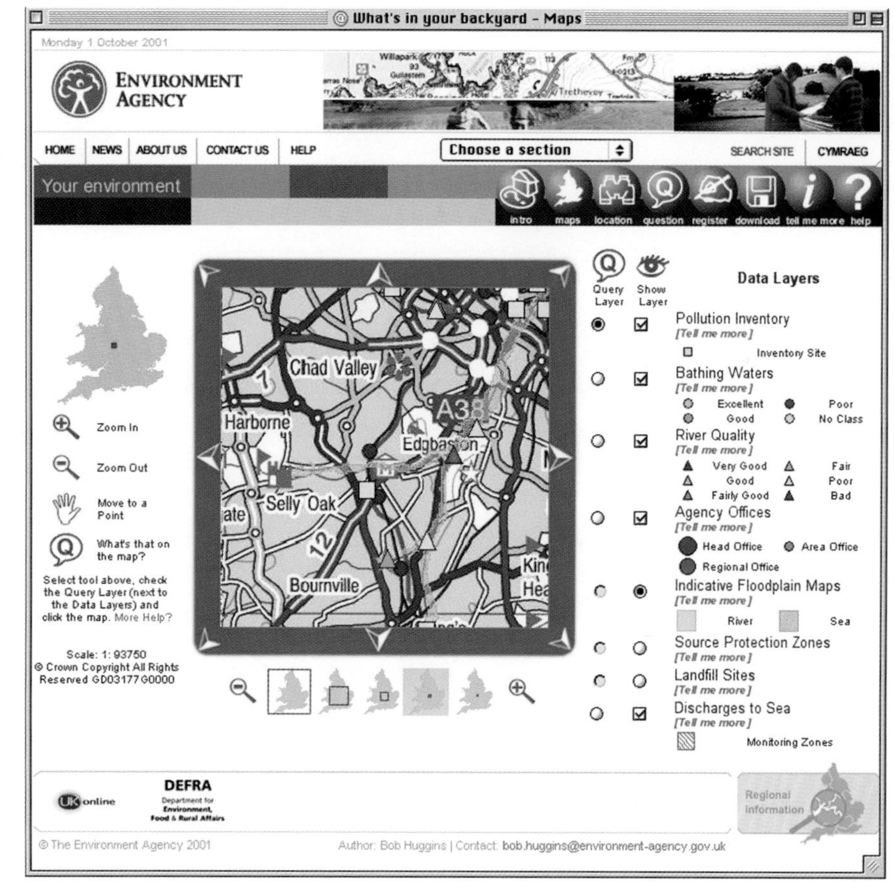

The Internet provides convenient access to environmental information from the Environment Agency. (www.environment-agency.gov.uk)

THE MERSEY BASIN – INTEGRATED MANAGEMENT IN ACTION

The River Mersey provides an example of large-scale integrated catchment management leading to high-quality redevelopment.

The catchment of the River Mersey includes a great deal of industrial north-west England. Near the mouth of the Mersey are Liverpool and Birkenhead and further upstream is the Manchester conurbation. The headwaters of tributaries of the Mersey reach up into the Lancashire mill towns, into the Peak District and into industrial Cheshire. Some 5 million people live in this area. The rivers and canals of the area together comprise some 2000 km of waterway.

The region has had a long industrial and trading history, originating in the water power available from the steep river valleys, and the natural harbour of Liverpool. The earliest industry was based on wool, followed by cotton. The abundant and accessible coal reserves of the area led to early industrialisation. Salt deposits in Cheshire, in the southern part of the basin, led to the early development of a chemical industry. Nowadays, petrochemicals, heavy chemicals and engineering manufacturing are still major industries in the area.

Not only did these industries produce large amounts of pollution that were conveniently discharged into the rivers of the area, but the people needed to work these industries produced large quantities of sewage, also discharged into the rivers.

This historic photograph (c. 1910) illustrates the severity of pollution produced by the alkali industry of the Mersey Basin.

In 1845 Engels wrote of the River Irk, a tributary of the Mersey

> ...a long string of the most disgusting, blackish-green slime pools are left standing on this bank, from the depths of which bubbles of miasmatic gas constantly arise and give forth a stench which is unendurable even on the bridge forty or fifty feet above the surface of the stream.

Over a 100 years later, things were little better – in the early 1980s the rivers of the Mersey Basin were acknowledged as the most polluted in Britain. Less than 50 per cent of the river length was capable of supporting fish, even including the streams in the rural uplands. The problems of sewage and chemical pollution remained, the latter made all the more obvious by the rise of the detergent industry and the attendant banks of foam at the many weirs.

To add to the woes of a severely polluted aquatic environment, changes in manufacturing and transport patterns resulted in the decline of the canals and the associated dock facilities, including the Manchester Ship Canal, so industrial dereliction became widespread.

The turning point came in 1985, when the government of the day took action, announcing the 25-year Mersey Basin campaign, with three objectives:

- to improve the quality of all rivers by 2010, so that all rivers, streams and canals are clean enough to support fish;
- to stimulate attractive waterside developments for business, recreation, housing, tourism and heritage;
- to encourage people living and working in the Mersey Basin to value and cherish their watercourses and water side environments.

The Campaign comprises a unique mixture of central government, large industry and commerce, and private individuals. Central government provides administrative support and grants through the Mersey Basin Campaign, worth around £415 000 per annum. Local industry, including many household names, supplies another £350 000 per annum in business sponsorship and support in kind, as well as bringing managerial skills to the project. The value of voluntary support provided by the Mersey Basin Trust, a registered charity, is valued at more than £400 000 per annum.

The rivers of the Mersey Basin, and their associated canals, drain an area of 4680 km^2, from Nantwich in Cheshire to Burnley in Lancashire. (Mersey Basin Campaign)

Armed with litter-pickers, a group of children prepare to collect rubbish. Clean rivers are the responsibility of everyone. (Mersey Basin Campaign)

The total investment necessary over the 25 year campaign period is estimated as £4 billion. Most of the necessary investment is needed to improve the quality of waste water treatment and is financed from charges on water used and sewage disposed. No longer does Liverpool discharge 950 million litres a day of untreated sewage into the Mersey. Wrong connections and combined sewer overflows on the tributaries of the Mersey are being tackled. Individual industries are investing to meet ever tougher standards

Foam on the River Mersey at Howley Weir, Warrington, produced by effluents from domestic detergent manufacture. (Mark Thewsey)

New developments at Salford Quays. Although much improved, water quality still needs remedial action. A bubbler (near the orange buoy) injects more dissolved oxygen into the water.

for the waste water they discharge. New, less polluting, manufacturing processes are being developed by the chemical industries. The Campaign makes a difference by providing a mechanism for a large number of cumulative small improvements, by ensuring that improvements are co-ordinated and by making sure that the benefits are realised from improvements.

The enormous task of cleaning all the rivers has been made more tractable by dividing it up into individual River Valley Initiatives, each with a paid co-ordinator. The individual Initiatives reflect the overall aims of the Campaign, but bring the problem closer to the people involved.

Education is a key component with a full-time liaison officer, to bring on a generation aware of the mistakes of the past, and of the measures they need to take to look after their environment. Involvement of volunteers of all ages in environmental projects builds a sense of ownership.

Fifteen years into the Campaign, benefits are showing. Eighty per cent of the watercourses now support fish. Significant waterside regeneration has occurred in Liverpool, Salford, Northwich, Stockport and Warrington, amongst others. The most notable has been at Salford where the docks have been rejuvenated. The award winning redevelopment of Salford Quays includes both housing and offices. A new tramway connects Salford Quays with Manchester City Centre. The most spectacular aspect of the regeneration of the docks is the recently opened Lowry Centre for the performing arts. The problems are not all solved – pollution levels are still sufficiently high that additional aeration is needed at times to maintain the environment suitable for fish. Green water indicates an excess of nutrients, but is a great improvement over the days when the water was black and lifeless. The achievements so far were recognised when the Mersey Basin Campaign was awarded the Riverprize at an international conference in Brisbane in 1999.

Financial support from many well-known companies has been important for the success of the Mersey Basin Campaign. (Mersey Basin Campaign, rights in trademarks reserved)

JOSEPH RANK LIMITED

5 Challenges for Urban Revival

Building sustainable communities is a key theme of the UK's sustainable development strategy. According to Lord Rogers' report *Towards an Urban Renaissance*, published by the Department of the Environment, Transport and the Regions in 1999, the most sustainable development option is to concentrate people, homes and jobs at the hearts of high-quality urban areas. This will reduce energy consumption by transport and avoid the further loss and degradation of the countryside. A well-designed urban fabric involves mixed and diverse urban areas strongly connected with one another through a network of sustainable transport options and green corridors. Attractive waterfronts and blue networks should be key elements of urban design. Clean rivers with a wide range of animals and plants have become symbols of a healthy environment, an attractive city, and a stakeholder society having ownership of its environment and responsibility for the needs of future generations.

At all levels of involvement in the urban revival process, progress is being made in tackling the range of key issues through new research and development, advanced planning, implementation of practical solutions to local problems, and better education. Within national and local government there is the adoption of new technology in building, transport, water management and energy recycling; in reducing the production of wastes; and in restoring natural areas within the built environment. Emerging grass-roots groups are also addressing local issues in practical ways.

BOX 5.1 **Planning to face the river in south-east England**

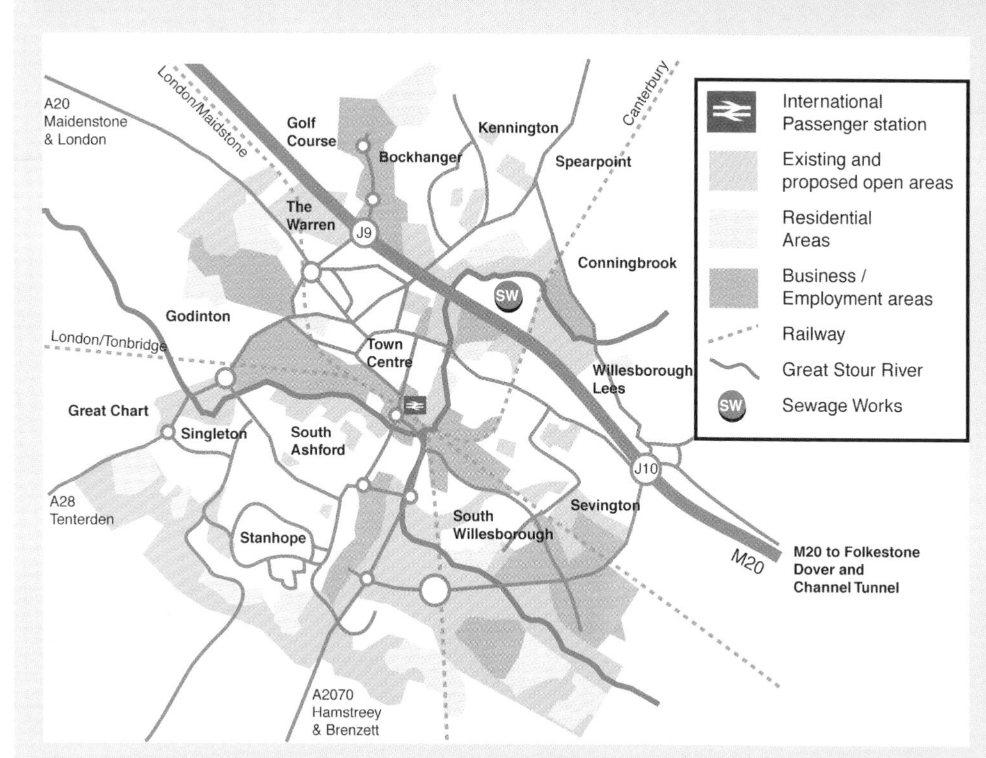

The Local Plan for Ashord, Kent, contends with the M20 motorway and the Channel Tunnel rail terminal, but still finds space for the river.

To accommodate projected housing needs for south-east England, current figures suggest that 310 000 houses should be built in the next 5 years alone. Ashford is one town that will be profoundly affected by this policy, given further impetus by the Channel Tunnel international rail terminal and the proximity of the M20. The local authority is already committed to the construction of 7983 new houses between 1996 and 2006, and it is likely to be asked to increase this figure substantially. Despite such development pressure, the Council has sought to maximise the environmental values of the Great and East Stour river corridors running through the town and to mitigate the effects of new building on these corridors.

The current proposed modifications to the Ashford Borough Local Plan explicitly state a presumption in favour of development adjacent to the corridors "provided that they also make a positive contribution to the function and amenity value of these corridors". In practice this has involved a commitment on behalf of potential developers to funding environmental improvements. Care has also been taken to design residential layouts with the riverfront being treated as positive space. Development is being re-orientated to face the river directly, rather than the river being hidden away behind back gardens. Vehicular access is to the rear of the houses; the front is reserved for footpaths and cycleways. It is a bold attempt to design high-quality space that is intended to be an asset to the community.

Prestige riverside development in Ashford, Kent.

Firstly, they are tackling the need for immediate clean-up to remove rubbish from their local streams and rivers. Secondly, they aim to improve education, to raise awareness and to foster ownership of local problems.

OPPORTUNITIES FOR REVIVAL

The demand for new development remains a key threat to the environment of the UK. This is true for land both outside and inside existing urban areas. In the current enthusiasm for an "urban renaissance" a target has been set for 60 per cent of new development to occupy reused brownfield sites by 2008. Planning Policy Guidance Note 3 (Housing) advocates densities of up to 50 houses per hectare, with a directive that brownfield land must be used before greenfield equivalents. This "compact city" will pose new challenges and opportunities for planners and developers, requiring as it does higher densities and sustainable, high-quality building. In addition, land must be found for industrial, commercial and retail development, together with improvements to the transport infrastructure.

Much progress has been made in "greening" cities in recent years, and there are many examples of creating nature areas on former derelict sites. However, there can be a potential conflict between the needs of the environment and the need to identify land for intensive development. Informal habitats evolving on disused urban sites, for

BOX 5.2 **Nature in the city**

One example of what can be achieved by redeveloping derelict land is to be found at Barn Elms on the south bank of the Thames between Hammersmith and Putney Bridges – in the heart of London. The Victorian reservoir site had supplied London with drinking water for 100 years and was known in the 1930s, a time when the Thames was still severely degraded, as the locus classicus of British birdwatching. In 1974 the site was designated a Site of Special Scientific Interest, due to the high number of wintering diving ducks attracted to the site.

Now owned by Thames Water, the site became redundant when the Thames Water ring main was completed. The reservoirs and water works have been converted into a 42-hectare wetland site – the largest wetland habitat creation project in Europe – in an innovative partnership between the Wildfowl and Wetlands Trust (WWT), Thames Water and the house builders Berkley Homes Ltd. Funds generated from the sale of the homes – some £11 million – were used to develop the basic landscaping and the construction of a visitor centre.

The centre, opened in May 2000, consists of three complementary areas: Peter Scott Visitor Centre building, named after the founder of WWT, exhibits, and the reserve. The buildings, which cluster around a central paved, open courtyard, consist of a theatre, where the Planet Water Show runs throughout every day, a glass observatory, an art gallery dedicated to the work of environmental and wildlife artists, and a discovery centre housing high-tech displays and exhibitors.

The former concrete water-supply reservoir at Barn Elms has been developed into a mosaic of wetlands providing habitat for a wide range of water birds. (Reproduced with kind permission of Thames Water)

The Wetland Centre is the first visitor attraction in London designed specifically to provide habitat for a diverse range of water birds, including many species that are rare or endangered within southern England. The emphasis on diversity is reflected in the range of habitats for feeding, roosting and breeding which have formed the final design, from open water lakes and reedbed, to seasonally inundated grass land and open mudflats.

example, may be at risk. It is important, therefore, that brownfield sites are assessed carefully and are not seen simply as an easy option for development. This note of caution is equally applicable to urban rivers. The waterfront has become an immensely saleable asset in new development, but it is essential that building work is sympathetic to its context.

Waterfront regeneration

Urban river corridors have been used as a catalyst for the regeneration of the wider urban area in a number of cases. The scope for redevelopment is immense. The decline of the major river ports and the associated industries of shipbuilding and engineering left their parent cities with increasing amounts of derelict, polluted land at the heart of the urban area. This land can, however, be successfully reclaimed.

The 121-hectare Greenwich Peninsula is one of Europe's biggest regeneration projects, with £180 million invested in site cleaning and preparation and infrastructure works. The project includes the Millennium Village, at the heart of which is a new artificial ecology park. A network of green corridors connects the village to the Thames river front. The scheme includes 1.2 hectares of created lakes and wetlands and innovative ecological flood barriers comprising newly planted terraces with reedbeds and saltmarshes. The scheme claims to have restored the stretch of the river close to the Thames Barrier to its past environmental glory with the reintroduction of a wide range of natural habitats lost by the industrialisation of the site. Wildlife such as shelduck, lapwings, ringed plovers, dunlin and redshank are expected to return.

Restoration of the Greenwich Peninsula includes a new futuristic urban development, the Millennium village, built around a man-made ecology park. (English Partnerships)

The Greenwich Peninsula is designed as a model for a society of the future. It rejects the single-function neighbourhood model and creates a new community consisting of a mixed use, residential and commercial area with leisure, shopping and recreational facilities all linked by a series of parks, corridors and transport links.

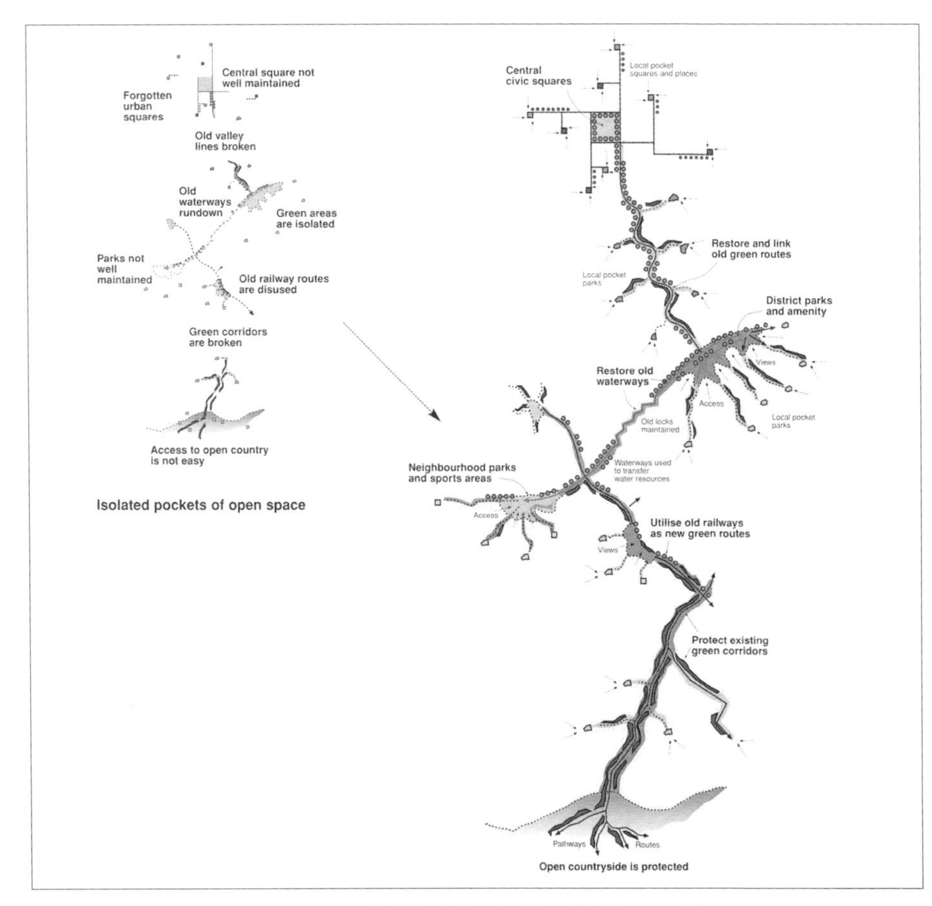

Urban areas should be designed to link residential areas, public space and natural green corridors. (Andrew Wright Associates)

Making the best use of river corridors

A "healthy" river is widely accepted as socially desirable; it is a reflection of the quality of life, and the quality of "the city". The urban process has led to the degradation of urban watercourses and their rehabilitation is central to the regeneration of the urban environment. Economic and social benefits will arise through environmental enhancement, improved public health, greater amenity and recreational opportunities, and increased competitiveness.

Rivers now have a place in many urban development and regeneration plans, but a co-ordinated strategy is necessary for the successful planning of urban rivers. A good example of where this principle has been applied successfully is Leicester. This scheme, led by the City Council, has secured extensive community involvement. A great deal of effort has been expended in education programmes and in encouraging recycling, transport efficiency and a variety of other policies.

The most tangible achievement of the scheme is the creation of a linear green corridor along the River Soar and the Grand Union Canal and, using the route of a former railway, the Great Central Way. The corridor traverses the entire urban area, connecting the city centre to the rural edge. Riverside parks, cycleways and footpaths have been established, largely on reclaimed derelict land, complementing the city's drive to reduce pollution and provide alternatives to car use. The green network is linked to the Watermead Ecology Park in the north of the city, a riverside nature reserve excavated from abandoned grazing land in 1982. Again, community involvement has been a vital element of the work. Volunteer groups assist with the maintenance of the sites. Particularly important is the role of Environ, a large environmental charity based in the city, which manages practical community projects. The success of the scheme has been recognised by its designation as the first Environment City in the country in 1990 and the receipt of a European Sustainable City Award in 1996.

Examples of new schemes founded on river corridor regeneration include the Stoke-on-Trent Rivers Strategy and the regeneration of the east side of London around the River Lea. Within the Lower Lea, local authorities, the Environment Agency, Thames Water, British Waterways and the Lower Lea Project, supported by European funds, are now actively involved in establishing sustainable environmental improvements as a catalyst for regenerating the heart of East London.

The Lower Lea includes the Newham "Arc of Opportunity" between Stratford and Docklands. The project offers an opportunity to restore a sense of the natural order of the Lea valley. The plan places emphasis on the creation of a healthy and sustainable river corridor within a vision of new urban water features to create a distinctive heart for the new metropolitan area. The rising groundwater in central London will be exploited by pumping, providing a supply of high-quality water to sustain newly created ecosystems within an 80-hectare "wet square" and canal system connecting the new Stratford town centre with the Thames. Urban runoff will be carefully managed using SUDS and rooftop collections to provide water for flushing WCs and watering gardens. These measures complement major infrastructure investments, including a new pumping station, storage and screening facilities, and installation of a fixed oxygen bubbler in the Lea Navigation, to protect the Lower River Lea from sewer discharges arising from equipment failure, sewer overflows during heavy rainfall and stagnant water.

At a larger scale, recent initiatives in the Thames corridor are significant in focusing specifically on the river as a basis for urban planning policy. The river is now covered by strategic planning guidance issued in 1997, with riparian authorities responsible for delimiting a "Thames Policy Area". There is an additional sub-regional planning framework for the Thames Gateway area, where environmental concerns have to be balanced with development pressure in growth areas along the estuary. The principal objectives of the strategic planning guidance are defined as being to:

- maintain and improve the quality of the built environment;
- restore and promote the vitality of the riverside in areas of development opportunity;
- conserve and enhance the character of the natural and historic environments; and
- encourage and facilitate the use of the river and riverside for transport and recreational purposes.

BOX 5.3 **The City of Stoke-on-Trent Rivers Strategy**

The City of Stoke-on-Trent Rivers Strategy, published in 1999, aims to reclaim the city's river corridors through a process of direct action and development control. A network of interlinked green corridors forms the structure for city development, keeping valley bottoms green and avoiding urban sprawl. The strategy seeks to utilise eight watercourses, as small as one metre in width, which are culverted or channelised and polluted for much of their courses through the city. The long-term goals are to open up the water corridor network to people who will be encouraged to take pride in their environment. This will be realised by providing opportunities for recreation and amenity along the network where wildlife can thrive, and where watercourses can fulfil their function of conveying and storing water in a way that enhances rather than detracts from the environment.

The Rivers Strategy policies actively seek improvements to riverside areas or zones which will be detailed in river action plans and the city plan. All development proposals close to watercourses will be fully assessed in terms of their effects on the aims of the River Strategy. Planning applications for development must ensure that the nature conservation and landscape value of the river corridor is safeguarded or enhanced. This will include the protection of sites of archaeological, cultural or recreational value; and ensure that the development relates to the waterside location. The natural character of the river channel must be maintained or restored and the integrity of the flood plain preserved. In addition, waterside access for pedestrians and cyclists, and for maintenance, should be safeguarded.

Longton Brook, Stoke on Trent. Rivers can be used as the focus of open spaces.

The Department for Transport, Local Government and the Regions recommends the Thames corridor strategic planning guidance as a good model for supplemental policy within the existing framework.

The role of science and technology

Developments in science and technology lead to new opportunities for the diagnosis and treatment of urban problems. Advances in water collection systems and sewage effluent treatment are creating opportunities for water re-use, reducing water stress in dry areas and improving the quality of rivers, enhancing opportunities for wildlife and recreation. These advances enable the implementation of new concepts for urban development.

One major national research programme – URGENT (Urban Regeneration and the Environment), funded by the Natural Environment Research Council – seeks to introduce a sustainable development regime that reshapes the structure and use of the urban environment. This important initiative was launched in 1997 to link academic researchers, city authorities, industry and regulatory bodies in addressing key issues such as the impact of rising groundwater, river flooding and ecological restoration.

The programme includes five projects on urban rivers. These include improved rainfall-runoff modelling and the development of new habitat and biological assessment methods for urban rivers. Studies will be carried out on disease burden levels attributable to combined sewer overflows, sediment (and sediment-associated contaminant) movement through urban areas, the ecological risks associated with urban river sediments, and the effects of dredging on water quality. The development of models will explore the effects of changes in flow regime, water quality, sediments and physical habitat upon animals and plants, and recreational uses.

Together the projects aim to provide the scientific basis for a new generation of tools and approaches to designing, planning and managing urban streams and rivers. The research will lead to the development of new methods to assess the actual and potential wildlife value of urban river reaches, to diagnose ecological problems, and to predict the effectiveness of new runoff and diffuse-pollution control technologies. The programme will not only lead to improved restoration and management strategies for streams and rivers within urban areas, but also create opportunities for the improved management of the water cycle within large catchments, and even across catchment boundaries.

An uncertain future

The 21st century is likely to witness climate change, but the effects this will have on weather patterns across the UK, on water resources and on the distributions of animals and plants, are uncertain. Nevertheless, risk and uncertainty are being incorporated into planning frameworks, for example,

BOX 5.4 **Regional opportunities for urban water re-use**

On a large scale, water resources derived from outside a region – and used and re-used several times before being discharged via a river or by direct pipeline to the sea – can also support growing demands. Today, about 1.8 million people live in Birmingham and the Black Country and their main supply of water is water-supply reservoirs in mid-Wales. The River Tame provides the primary vehicle for disposal of treated waste waters.

Discharges of treated sewage effluent from Minworth and the nearby Coleshill works produce a combined dry weather flow of about 550 million litres per day. In summer 80 per cent of this flow may be made up of treated sewage effluent. As the Tame enters the River Trent, these discharges have more than doubled the dry-weather flow in the main river. Improved sewage treatment has created opportunities for the re-use of the water after it has passed downstream and been subject to natural self-purification processes.

The Trent-Witham-Ancholme scheme was established in the 1970s and is licensed to transfer up to 182 million litres per day from the lower Trent at Torksey to meet water demands for agriculture, industry and public water supply in North Lincolnshire. More recently the Trent has been identified as a source of future supply to meet water needs in the East Midlands. A new abstraction on the middle river near Shardlow incorporating worked-out gravel pits will provide bankside storage. To protect existing uses of the Trent, such as navigation and effluent dilution, new licences are subject to a prescribed flow below which abstractions must reduce or cease.

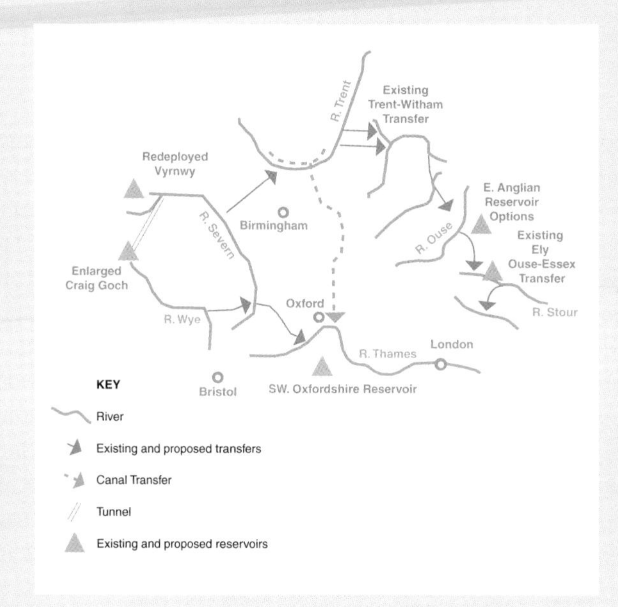

Options for inter-basin transfer to support demand for water supply.

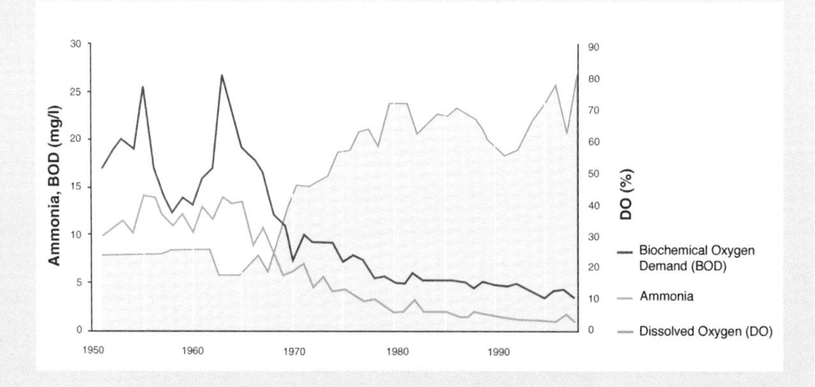

Increasing dissolved oxygen, and decreasing concentrations of the pollutants ammonia and BOD, indicate the improvement in water quality in the River Tame.

The discharge from Minworth forms a major part of the flow in the River Tame (the froth is a consequence of the large discharge, rather than a sign of pollution).

for large projects such as reservoirs, and for diverting local government planning for housing to areas less prone to flooding.

Globally, surface temperatures are likely to increase, which is likely to lead to a more vigorous hydrological cycle, with more extreme events – floods and droughts – in many areas. In the UK, average temperatures have increased by around 1 °C in the last hundred years. Computer models of the atmosphere predict that a further temperature increase of +1.2 to +3.4 °C is likely in the UK by 2080. They also predict that warming will cause generally drier summers, especially in the south-east, and wetter winters, especially in the north-west. The unusually hot dry summer of 1995 could become typical. Storms are likely to become more frequent.

There are immediate implications for urban rivers from these changes. Flooding incidents are likely to become more frequent. In the past flood risks were managed by flood warning, defence and emergency response. Although these remain important tools, in the future the joint consideration of land use and increased flood risk will be essential to avoid putting properties and lives at risk. The sustainable approach to flooding is to avoid building on river flood plains. Sea level will also rise, both as a consequence of ice-cap melting and because warmer oceans will expand. The anticipated rise in sea level in southern and eastern England is more than half a metre. This rise will affect the water level in the lower reaches of rivers draining to the sea, and will cause river levels to rise also, further increasing the risk of river flooding.

BOX 5.5 **Too little too late?**

Greenhouse gases, principally carbon dioxide and methane, trap infrared radiation emitted by the earth's surface, causing the temperature in the atmosphere to rise. Human activities, such as the burning of fossil fuels, industrial operations and forest clearing, enhance this natural greenhouse effect. It is widely held that the emission of these gases is a major contributor to observed climate change.

It is estimated that in order to stabilise current concentrations of greenhouse gases in the atmosphere, large cuts in emissions would be required. In June 1992 representatives of 162 governments signed The Framework Convention of Climate Change (FCCC) calling for a precautionary approach towards the threat of global warming. At the Kyoto Conference in 1997, the UK undertook to reduce carbon dioxide emissions to 80 per cent of 1990 levels by 2020.

Substantial environmental, social and economic costs of adapting to climate impacts in the future could be reduced by action to reduce emissions now. Several important approaches to reducing greenhouse gases include supplying energy more efficiently, generating energy from less polluting fuels and increasing our use of renewable energy. The simplest of all is to avoid car use for short journeys, and to drive within speed limits. Current technology can also offer simple ways to reduce emissions, such as better insulation in buildings, energy-saving electrical appliances and cars which use less fuel.

However, research has shown that the climate system will take many decades to respond to the reduction in carbon dioxide emissions. In designing urban areas it is important to plan flexibility to allow for adaptation to climate change, to incorporate climate change into scientific models to test future scenarios, and to implement policies that work with changing climate.

Actions are already being taken to protect against higher sea levels in some areas. In other areas, there is no sustainable choice other than to let the sea reclaim land. Improved flood warning and a new culture of wise water use will be required as our environment becomes a more hazardous place to live, with floods and droughts becoming increasingly common.

Measured average summer temperatures in southern England for 1961–1990 and the temperatures predicted for the 2080s by the UK Climate Impacts Programme (medium high scenario).

The Thames Barrier helps protect London from tidal flooding. (Reproduced with kind permission of Thames Water)

Increasing demand for water in the dry south-east of England may result in frequent summer water shortages and reduce summer river flows below ecologically acceptable amounts. Increased water stress may result from drier, more predictable summers attracting more visitors, increased holiday or second home ownership, and people choosing to return to the south-east on retirement. However, increased irrigation for agriculture may cause the greatest problems. Gardens will evolve with low water requirements and the concept of the zero-water garden may be advanced with grasses, hardy palms and extensive use of gravels and pebbles to prevent evaporation from the soil, and water coming from roof runoff collected in water butts.

Planning for the effects of climate change will need to take into account new information and the increasing rate of information generation. Monitoring of changes in rivers is likely to play an increasingly important role.

The need to manage water supplies for rapidly expanding urban communities is a global issue:
Riverside slums in Ho Chi Minh City, Vietnam. (Axiom Photographic Agency)
Riverside in Chicago, USA. (Geoff Petts)

THE INTERNATIONAL PERSPECTIVE

Over the past decade a number of international symposia have exposed major problems for the planet in this 21st century. These problems arise from increasing population growth and the concentration of people in large cities, and are linked to the challenges of freshwater management and the fight against poverty. In many areas of the world, the number of water taps per 100 persons is a better indicator of human health than the number of hospital beds. At the beginning of the third millennium there are 1.3 billion people without safe drinking water and 1.9 billion lacking appropriate sanitation. There are 30 or so major diseases in the developing world, and of these 21 are water and sanitation related. Floods, too, are a major problem in many regions.

Globally, with an urban population expected to reach 6 billion by 2025, growing cities will place increasing demands upon surface water and groundwater supplies and there is the danger of increasing water contamination. Where rivers cross international boundaries, conflicts over the management of the catchment can lead to major international disputes and there are many who predict that future wars will be fought over water issues. The importance of this is has been recognised internationally for some time, and the United Nations has taken a leading role in promoting catchment-based management, particularly in relation to trans-boundary pollution.

A busy quayside, Motlava, Poland. (John Pomfret)

Water is fundamental to so many aspects of life and of the environment, and cuts across so many areas relating to sustainable development, that the management of urban waters poses considerable challenges. What is clear is that sustainable solutions will require institutions and individuals to work collectively through local, national and international strategies. Across Europe, for example, programmes are being advanced to encourage the sharing of best practice.

The United Nations Conference on Environment and Development in Rio de Janeiro in 1992 established a blueprint for future survival on planet Earth – now known as Agenda 21. The key principles for integrated water management, as cited in Agenda 21, are:

- water is a scarce resource;
- all those who are interested in water allocation and use should be involved in decision making;
- water should be managed within a comprehensive framework, including water supply and waste management.

Only in this way will we ensure that what we do today does not undermine the development and environment needs of present and future generations.

In 1997 UNESCO convened an international conference in Paris entitled "Water, the City and Urban Planning". Delegates from 42 cities and 50 countries compared their visions of a world in which the great majority of the population will be city dwellers. In a world where the irreversible march of urbanisation is causing increasing pressures on water resources and where the growing sanitary and ecological problems arising from the human concentrations and disposal of wastes pose formidable challenges, they considered that:

- sustainable urban development threatens to become an elusive goal;
- the duality of water as an endowment and an economic good must be faced with wisdom and compassion;
- a responsive population with proper mechanisms to participate in decision making at local and higher levels is essential for effective results.

BOX 5.6 **Water City International**

Launched in April 1999, Water City International is a joint-venture project to promote planning within cities in the North Sea region, emphasising all aspects of water in the city. The project includes the cities of Leeuwarden in the Netherlands, Göteborg in Sweden, Emden in Germany and Norwich in the UK. The target areas for the project include:

- preserving, protecting and improving water quality and quantity;
- using water to develop the natural environment within the city;
- reviewing the recreational use of water;
- considering the role of water transport;
- integrating water in urban planning;
- highlighting the use of water as a tool for improving urban living;
- seeking to respond to opportunities created by technological developments in water purification, channel restoration, and sustainable energy.

Restored riverside at Norwich provides opportunities for development and recreation.

BOX 5.7 **The Landcare movement in Australia**

The Landcare movement in Australia was established in the 1980s to empower community groups to participate in the conservation of their local environment. Members of the community form a Landcare Group to identify specific environmental problems and projects to tackle them. Each region or district is allocated funds by the national Government and, upon consultation with the Department for Land and Water Conservation, a group may apply for funding to resource each project. As the projects are predominantly "soft" engineering, with abundant free labour at hand within the group, funding is generally for small amounts, for example, to provide seedlings or the use of a piece of machinery for a day.

Throughout the late 1980s the Landcare movement spread rapidly throughout the entire country. The success of these small community-driven groups encouraged the Government to provide more money towards the schemes, so much so that the government christened the 1990s "The Decade of Landcare".

As Landcare expanded, the focus of groups became more specific. Soon the names Rivercare, Dunecare, Saltcare and Bushcare became recognisable offshoots. Urban Landcare took hold in the cities. The diversity of groups and their objectives, and the increasing community interest, led to the formation of another type of community group to oversee the bigger picture. Total Catchment Management (TCM) was established to oversee the conservation strategies of each major river catchment. Landcare groups continued to do the on-the-ground work. The TCM committees provide the overall vision for land management within the catchment, oversee community education, prioritise projects and issues and liaise directly with all levels of government.

The Government estimates that Landcare saved it billions of dollars during the 1990s. More importantly, the environment has been conserved in a way impossible to quantify in monetary terms.

Throsby Creek is a small urbanised catchment in Newcastle, an industrial city 160 km north of Sydney, with a population of 67 000 people. The catchment area is only 3000 hectares and 77 per cent of it is classified as built up. But the Throsby Creek Total Catchment Management (TCM) Committee, established in 1987, recently won the 2000 Gold Award for Landcare/TCM. In the committee's own words, they "took a dirty, smelly drain, where nearby residents built high fences to block it out, and turned their properties into something more valuable with a waterfront where families now picnic, cycle, and clean up litter and plant trees".

Central to the project was the removal of concrete drains in key areas and the planting of local native riparian vegetation species throughout the catchment. As well as the people, the native vegetation, the waterbirds and the fish are returning to Throsby Creek. The committee believes that their ongoing success is more than about just getting funding and doing hands-on work. Two key factors were the co-ordination of the activities by the Landcare groups and their education programmes that have increased and, just as importantly, maintained environmental awareness and enthusiasm within the community.

The Paris Statement recognised the existence of a wealth of experience that must be shared in order to learn how it may benefit other cities, but also that each city has a set of particular conditions and problems. Special problems arise in coastal cities; by 2025 some 75 per cent of the world's population is predicted to be living within 60 km of the sea.

Lessons from developed economies suggest that sustainable urban development can be achieved through planning policies that seek to restore and maintain the status and vitality of "blue arteries" by building on five important functions that rivers fulfil:

- drainage and water supply;
- open spaces and ecological corridors;
- transport networks;
- recreational, leisure and tourist facilities;
- a setting for and access to both new development and heritage sites.

Vacant land, neglected open space, unmanaged edges and obsolete structures create new opportunities. The river corridor must be seen as a strategic open space. This position is strengthened by the value of river frontage, and of open vistas and skylines; if the space is a well-connected network that links rivers with other green corridors to improve public access, it is also enhanced by policies promoting the river as a focus for water-based activities. With foresight, governments, industry, researchers and the public can come together to define this common agenda.

The River*prize*, awarded in Brisbane, Australia to the Mersey Basin Basin Campaign in 2000.

BIBLIOGRAPHY

Towards an Urban Renaissance: final report of the Urban Task Force chaired by Lord Rogers of Riverside, London, Department of Transport, Environment and the Regions, 1999.

Quality of Life Counts: Indicators for a strategy for sustainable development for the United Kingdom - a baseline assessment, London, Department of Transport, Environment and the Regions, 1999.

CIRIA, *Sustainable urban drainage systems – design manual for England and Wales*, CIRIA C522, 2000.

Environment Agency, *Environment 2000 and beyond*, Environment Agency, Bristol.

Environment Agency, *River Habitat Quality, the physical character of rivers and streams in the UK and Isle of Man*, Environment Agency, Bristol, 1998.

Scottish Environment Protection Agency, *Watercourses in the community*, Scottish Environment Protection Agency, Stirling, 2000.

Trench, R. and Hillman, E., *London under London: A subterranean guide.* John Murray (Publishers) Ltd, London, 1984.

United Nations Conference on Environment and Development, *Agenda 21: programme of action for sustainable development: the final text of agreements negotiated by governments at the United Nations Conference on Environment and Development (UNCED), 3–14 June, 1992, Rio de Janeiro, Brazil*, United Nations, New York, 1993.

Ward, D., Holmes, N. and Jose, P. (Eds), *The new rivers and wildlife handbook*, RSPB, Sandy, 1994.

Roberts, P. and Sykes, H. (Eds), *Urban Regeneration: A Handbook*, Sage Publications, London, 2000.

Calow, O. and Petts, G.E. (Eds), *The Rivers Handbook*, 2 vols, Blackwell Scientific, Oxford, 1994.

Boon, P.J., Davies, B.R. and Petts, G.E. (Eds), *Global perspectives on river conservation: science, policy and practice*, Wiley, Chichester, 2000.

Downing RA, *Groundwater – our hidden asset*, British Geological Survey, 1998.

Web Sites

Environment Agency: http://www.environment-agency.gov.uk/

Environment and Heritage Service, Northern Ireland: http://www.ehsni.gov.uk/

European Rivers Network (Rivernet): http://www.rivernet.org/

River and Ocean Research and Education: http://www.rore.org.uk/

River Restoration Centre: http://www.aecw.demon.co.uk/rrc/rrc.htm

Scottish Environment Protection Agency: http://www.sepa.org.uk/

UK Department of the Environment, Food and Rural Affairs: http://www.defra.gov.uk/

UK Department for Transport, Local Government and the Regions: http://www.dtlr.gov.uk/

UK Rivers Network: http://www.ukrivers.net/

Water City International: http://www.watercity.org/project.html

Water UK: http://www.water.org.uk/

WWF European Freshwater Programme: http://www.panda.org/europe/freshwater/

River Thames at Battersea.
(John Barnes Design)